Agricultural Waste - New Insights

Edited by Fiaz Ahmad
and Muhammad Sultan

Published in London, United Kingdom

Agricultural Waste - New Insights
http://dx.doi.org/10.5772/intechopen.100885
Edited by Fiaz Ahmad and Muhammad Sultan

Contributors
Mohd Muzamil, Sehreen Rasool, Ummyiah H. Masoodi, Kawaljeet Kaur, Preetpal Singh, Sharda Dhadse, Fiaz Ahmad, Aftab Khaliq, Muhammad Sultan, Ding Qishuo, Muhammad Ramzan Sheikh, Asma Aslam, Neelam Asghar Ali, Stephen Nnaemeka Okey, Cosmas Chikezie Ogbu, Arjun Singh, Saroj Choudhary, Rajendra Prasad Meena, Anchal Dass, Renuka Vinothkumar, Frank Lucien, Janet Paterson

First published in London, United Kingdom, 2023 by IntechOpen
IntechOpen is the global imprint of INTECHOPEN LIMITED, registered in England and Wales, registration number: 11086078, 5 Princes Gate Court, London, SW7 2QJ, United Kingdom

British Library Cataloguing-in-Publication Data
A catalogue record for this book is available from the British Library

Additional hard and PDF copies can be obtained from orders@intechopen.com

Agricultural Waste - New Insights
Edited by Fiaz Ahmad and Muhammad Sultan
p. cm.
Print ISBN 978-1-80356-965-9
Online ISBN 978-1-80356-966-6
eBook (PDF) ISBN 978-1-80356-967-3

For EU product safety concerns:
IN TECH d.o.o., Prolaz Marije Krucifikse Kozulić 3, 51000 Rijeka, Croatia,
info@intechopen.com or visit our website at intechopen.com.

Meet the editors

Dr. Fiaz Ahmad is an associate professor in the Department of Agricultural Engineering, Bahauddin Zakariya University, Multan, Pakistan. He obtained a BSc and MSc in Agricultural Engineering from the University of Agriculture, Faisalabad, Pakistan, in 2004 and 2007, respectively. He obtained his Ph.D. in Agricultural Bioenvironmental and Energy Engineering from Nanjing Agriculture University, China, in 2015. He completed a Postdoc in Agricultural Engineering from Jiangsu University, Zhenjiang, China in 2020. He was awarded a fellowship from the Higher Education Commission of Pakistan for Ph.D. studies and from the Chinese Government for post-doctoral studies. He is the author of more than sixty-five journal and conference articles. He also edited two books and is an editorial board member of several well-renowned journals. He has supervised ten master's students to date, and he is currently supervising four master's and two doctoral students. Dr. Ahmad completed three research projects and is currently working on two others. His research interests include various aspects of agricultural and farm mechanization.

Dr. Muhammad Sultan is an associate professor in the Department of Agricultural Engineering, Bahauddin Zakariya University, Multan, Pakistan. He obtained a BSc and MSc in Agricultural Engineering from the University of Agriculture, Faisalabad, Pakistan, in 2008 and 2010, respectively. He obtained a Ph.D. in Energy and Environmental Engineering from Kyushu University, Japan, in 2015. He completed post-doctoral research in energy and environmental engineering at the same university in 2017. He also did postdoctoral research in mechatronic systems engineering at Simon Fraser University, Canada, in 2019. He has published more than 300 articles in international journals, conferences, books, and book chapters. He has been a reviewer for more than 100 renowned journals and serves as editor for five others. His research focuses on developing energy-efficient temperature and humidity control systems for agricultural applications including greenhouse, fruit/vegetable storage, livestock, and poultry applications. His research includes adsorption heat pumps, desiccant air conditioning, evaporative cooling, the Maisotsenko cycle, adsorption desalination, energy recovery ventilators, atmospheric water harvesting, and wastewater treatment.

Contents

Preface

The increasing demand of agricultural production for human, animal, and industrial requirements is responsible for the enhancement of agricultural and agro-industrial activities. Each step of such activities produces various types of agricultural waste, including crop residue, on-farm livestock and fisheries waste, forest waste, agro-industrial waste, and so on. Currently, handling and managing agricultural waste are challenging tasks worldwide, especially in the context of environmental pollution control and sustainable agriculture. Thus, efficient management in terms of reuse, recycling, and reduction of agricultural waste is principally needed not only for the green economy but also for farmers' profitability. This will also contribute to minimizing environmental pollution, greenhouse gas emissions, and climate change in order to meet the 2030 United Nations Sustainable Development Goals.

This book, *Agricultural Waste – New Insights*, focuses on agricultural waste production and management in the multidimensional aspects of crop residue, biodegradables, biomass, composting and vermiculture, agricultural waste economics, air pollution, environmental safety, waste management and handling, on-farm waste reuse, and agricultural waste value addition. It provides a comprehensive review of recent developments in agricultural waste management and handling.

Agricultural Waste – New Insights includes nine chapters that present novel concepts, new insights, and opportunities related to agricultural waste management.

Chapter 1, "Crop Residue Collection and Handing Machinery Performance: A Review", provides a review of crop residue collection and handling machinery.

Chapter 2, "Food Wastage Footprint, Food Security, Environment and Economic Growth Nexus in Developing Countries", provides a worldwide account of the environmental footprint of food wastage as well as food security, focusing on impacts on environmental quality and economics.

Chapter 3, "*In Situ* and *Ex Situ* Agricultural Waste Management System", addresses the in situ and ex situ agricultural waste management system.

Chapter 4, "Nutrient Supplying Potential of Crop Residues in Indian Agriculture" discusses residue recycling for the circular nutrients economy.

Chapter 5, "Agro-Industrial Waste Management: The Circular and Bioeconomic Perspective", discusses global production and possible valorization of recyclable agro-industrial residues and food wastes.

Chapter 6, "Crop Residue Burning in India: Potential Solutions", examines issues of crop residue burning and potential solutions in India.

Chapter 7 "Recovery and Characterization of Astaxanthin Complex from Prawn Waste Extract and Its Separation Using Antisolvent Precipitation Technique" discusses the recovery and characterization of one of the major valuable components of prawn waste called the 'astaxanthin complex'.

Chapter 8, "Activated Charcoal: A Novel Utility Product for Enhanced Animal Health and Production from Agricultural Wastes (Pig Dung and Palm Oil Wastes)", presents a study evaluating the physico-chemical properties of activated charcoal produced from a blend of agro-wastes for use as feed additives.

Chapter 9, "Vermiconversion of Textile Industrial Sludge: Waste Management and Nutrients Recycling", discusses vermi-conversion of textile industrial sludge into vermicompost.

The editors are pleased to share the emerging research on agricultural waste management technologies, strategies, and applications. *Agricultural Wastes – New Insights* is useful not only for agricultural and environmental engineering professionals but also for associated agricultural scientists. Finally, we would like to thank the authors for their valuable contributions to this book.

Fiaz Ahmad and Muhammad Sultan
Department of Agricultural Engineering,
Bahauddin Zakariya University,
Multan, Pakistan

Chapter 1

Crop Residue Collection and Handing Machinery Performance: A Review

Fiaz Ahmad, Aftab Khaliq, Ding Qishuo and Muhammad Sultan

Abstract

Increasing demand of agricultural production for human, animal, and industrial requirements is responsible for the enhancement of agricultural and agro-industrial activities. Each step of such activities produces various types of agricultural waste that include crop residue, on-farm livestock and fisheries waste, forest waste, agro-industrial waste, etc. Currently, handling and managing agricultural waste is a challenging task worldwide, especially in the context of environmental pollution control and sustainable agriculture. Thus, efficient management in terms of reuse, recycling, and reduction of agricultural waste is needed not only for the sustainable agriculture but also for farmers' profitability. Various type of farm machinery is available and are in use to collect the crop residue from the field or directly incorporate the residue into the soil. The incorporated crop residue not only increases the soil fertility but also decreases the greenhouse gases emission due to burning of the crop residue. The crop residue chopper can be a solution of residue management at farmer field level. This chapter provides a review on the crop residue collection handing and incorporation machinery performance and their advancement.

Keywords: crop residue, collection, incorporation, agriculture production, sustainable agriculture, machinery

1. Introduction

Crop residue also known as the plant biomass or remnants of crops are natural resources that can be utilized by the farmers for the variety of purpose such as mulching, compositing, as animals feed, rural resident construction, and as energy resource at domestic and industrial level [1]. However, a significant percentage of the crop residue is burnt in the field with the objective to clear the field from stubbles and straw/stalk for time sowing of the next crop [2]. Usually, these crop remnants have a significant chance of refreshing the soil with a substantial amount of plant nutrients. Further, mulching of crop residue maintains the soil temperature, retains soil moisture and mitigate the carbon emission. Crop residue management practice is advised for preserving the natural resources and enhancement of crop productivity [3]. Using modern farm equipment such as combine harvesters, rotavators, and seed drills can lead to increase in crop residue production, primarily from cereals (74%), followed by sugar crops (10%), legumes (8%), tubers (5%), and oilseeds (3%). This ratio of crop residue production shows that the residue production increases with use of the advanced technology [4]. Typically, crop residues are burned in the fields worldwide, particularly in developing countries. This practice not only contributes to air pollution, but also impedes nutrient recycling and negatively impacts soil health by reducing microbial activity and causing carbon loss.

Collecting and handling of agricultural residues from the field is not only an energy- and labor-intensive process, but it is also time-consuming and can delay the sowing of the next crop, ultimately affecting crop yield [5]. Residue burning is not considered a best option to handle the crop residue in these days, as it also releases the greenhouse gases (GHGs) which are injurious to the environment and are a source of global warming [6, 7]. Therefore, effective crop residue management is required through eco-friendly methods to maintain the organic matter and nutrients recycling in the soil.

Therefore, the crop residues handling and management machinery get attention to control the soil degradation, environmental pollution, and to improve the soil health and crop productivity [8]. However, as bio-energy point of view the crop residues would be collected to use as bio-fuel in industry rather than incorporated into the soil, which is in contrast to the concept of conservation agriculture. However, it is crucial to streamline the processes for harvesting, threshing, and transportation, and also to the select varieties with desirable straw properties for bio-fuel generation [6]. Rice crop husk and straw can be used efficiently for the generation of bio-energy. According to empirical estimations, 1 tonne of rice husk has the ability to create between 410 and 570 kWh of electricity, and 290 kg of rice straw can produce 100 kWh [9]. In the rice-growing tropical continent, bio-gas generation from waste combined with animal excrement is an old method. However, its potency and utilization have decreased recently, it still has a lot of benefits, such as enhancing the soil fertility for crop growth [10]. The burning of crops' residues adversely affects the regional climate, production yields, and the human health [11]. The wide collection of studies on crop residue management's impact on crop productivity, soil health, environment protection, and energy potential has been made possible by the development of machines for shredding crop residues.

Garg [12] conducted a study to developed paddy straw chopper-cum- spreader. Jia et al., [13] developed a combined stalk– stubble breaking and mulching machine to solve the issues of existing stalk-breaking and stubble- breaking machines. Wang [14] design and developed a straw side spreading and no-till soybean seeding machine to seed soybean in wheat stubble. Elfatih et al., [15] conducted a study to evaluated the performance of the modified rice straw chopper for composting. A performance evaluation study of tractor operated mulcher for paddy straw was conducted by Verma et al., [16]. The effect of residue management of GHGs emissions have been investigated by Lehtinen et al. [17]; Wegner et al., [18]; Zhang et al., [19], on soil health estimated by Bisen and Rahangdale [20]; Clay et al., [21]; Turmel et al., [22]; Yadav and Arora [23]; Zahid et al., [24], crop yield quantification by Hiel et al. [25]; Koga and Tsuji [26]; Paul et al. [27]; Piccoli et al., [28], and as bio-energy product used by Ahiduzzaman et al., [29]; Chauhan et al. [30]; Devi et al., [31]; Fazio and Barbanti [32].

The main objective of this chapter is to present a comprehensively review on crop residue handling and management machinery performance which were designed and used in the field to enhance the crop productivity, environmental pollution control and soil health. This review article is organized in the following manner. First, we discussed the crop residue handling machinery development for different crop trash and then effect of management were explained on soil health, crop production and environment protection. Finally, we summed up the best trash management machinery performance for efficient crop residue management.

2. Crop wise residue handling machinery

2.1 Rice residue management

Rice (*Oryza sativa L.*) is the most significant crop and considered as the lifeline in Asia. Rice is the most grown food by the people in Asia and it is one of the most valuable and productive one in agroecosystem. The two main tends that evolved the rice cropping in various agroecological regions are intensification and diversification [33]. Limited delay between to improve the crop production to land usage intensification makes it harder to manage these residues, which can affect tillage and sowing procedures for the subsequent crop. However, the Asian farmers are more often burning agricultural remnants as a form of disposal, due to lack of the mechanized and limited technology for the management of huge quantities of residue [34]. Different resource conserving technologies i.e., direct seeding, minimum tillage, bed planting, crop diversification, and proper residue management are potential alternative options to reduce the energy input.

Paddy straw is the valuable asset for the farmers [35]. Therefore, the much needed to manage the paddy residue in the field. Many studies have been conducted about the rice trash management and handling machinery for soil health, crop productivity, and environmental safety control as shown in the **Figure 1**. Hegazy et al., [37] developed a zero-tillage seed drill which directly seed wheat seed after rice straw-chopping. Two motorized rotors make the straw chopper; the function of the first chopper is to chop the residue and then other removes it from the furrow openers. The straw chopper is assembled in front of the direct seed drill frame. The developed seed drill had field efficiency that varies from 1.89 to 1.94 hm^{2}/hr. and fuel consumption from 10.88 to 11.6 L/hm^2 with <11 cm straw chopping efficiency. Sidhu et al., [38]. developed a combo happy seeder that performs the harvesting and drilling process in single pass. The machine is manufactured in such as that it has a narrow strip tillage assembly in front of the sowing tines which consequently improves seed-soil contact on the sandy loam and loam soils. Regatti Venkat, [39] conducting the research study for rise residue management and conclude that the happy seeder technology is the significant solution for the management of crop remnants, and good option for direct seeding of wheat after the harvesting of paddy crop. Kathpalia et al., [40] was conducting the research study to investigate the impact of the crop residue management with happy seeder on the crop productivity enhancement.

2.2 Wheat straw management

Globally, approximated 225 Mha area is used to sow the wheat with the production of almost 684 Mt./y. [41]. The increase in sowing area enlarges the on field residue. Scarlat et al., [42] reported that the wheat cropping is one of the most invaluable source of biomass i.e., including wheat husk, mainly as straw, suitable for livestock as

Figure 1.
Rice residue handling and management machinery [36].

bedding, or as raw material for chemical applications. The authors found that the ~17% (w/w) is the production of the chaff as compared to wheat. Therefore, 138 Mt. of the spelt is harvested annually in Europe [41]. In addition, some proportion of the biomass is removed from the soil and does not play any role so the soil organic carbon is obtained from the decomposition of the root in the soil which is remained unaffected [43]. In fact, the remaining proportion of the plant/crop residues in the soil can help to conserve soil fertility and sediment. However, research has indicated that such effect may be widely diverse [44]. Additionally, it was clearly mentioned in the literature that minimizing soil tillage is more likely to decrease soil fertility and loss of soil organic carbon than maintaining the soil cover [45]. Recently, the combine harvesters had lack chaff recovery systems unless appropriate equipment has been installed, that effect the potential of the chaff for variety of purposes [46]. The potential use of agricultural waste products, such as wheat chaff, for non-food applications has encouraged the development of innovative methods by machine manufacturers. While some systems are currently available, others are still in the prototype stage [47]. The standard approach for chaff recovery systems, regardless of the brand, is to collect the chaff at the bottom of the combine harvester's cleaning shoe system before it falls to the ground and gets lost. When crop yields are low, residue collecting may not be both economically and environmentally feasible [48]. Collecting the chaff separately is the cost ineffective method including the higher cost of the transportation if shipping it without the balling operation [49].

Recently, in Pakistan the combine harvester has adopted at medium and large size farms because it reduces the harvesting time and cost, and grain losses. In Pakistan over 5000 units of combine harvester are working for the harvesting of wheat and rice. The field capacity of the harvester is much higher than that of other harvesting equipment. The limitation of the working of the combine harvester is it works only in the anchored high stubble crop. To mitigate this farmer are demanding the combine harvester which can cut the wheat stubbles and loose straw and chop the residue as fodder i.e., cattle feeding. For this purpose, wheat straw handling and management machinery get attention for the industrial and academic sector to develop and evaluate this machinery.

Zhang et al., [50] developed a stubble chopping shredder machine comprised on different components i.e., cultivator blade, depth cylinder, chopping blade rotor. Mahmood et al., [51] conducted the study to evaluate the performance of wheat straw in field as show in **Figure 2**. The study concluded that wheat straw chopper is effective and financially profitable technology for the farmers which saves the stubbles for the cattle feed. Suardi et al., [52] compared the two-wheat straw-chopping technologies. The study investigated that by setting the chaff collector with the combine harvester can benefit the farmer.

2.3 Sugarcane trash management

Sugarcane crop plays a significant role in the economy of Pakistan with collaboration of industrial sector and framers. It is the second largest cash crop of Pakistan [53]. The developing countries adopted manual method for sugarcane harvesting due to lake of mechanization. Sugarcane trash is becoming more available product in the field due to mechanical harvesting. However, due to the lack of labour and limited time of sowing for the next crop the sugarcane trash mostly burns on the site that produces toxic smoke. This show the need of mechanized policies for the timely sowing of next crop [54]. In Pakistan, the sugarcane trash management is the main problems arise after the harvesting of sugarcane. After manual harvesting, the trash management in the field is very difficult task for the farmers, so they burn it into the field. One of the significant concerns during sugarcane trash burning in the field is the environmental pollution and damage to soil microbes. The large amount of carbon, nitrogen, and particulate matter (PM) are the major chemical pollutants that are emitted during the trash burning. These emitted pollutants have severe effects on the composition and acidity of rainwater. Further, the emissions of mister and trace gases from the trash burning are very effective for human health [55]. Therefore, to mitigate the environmental pollution and enhance soil fertility worldwide, sugarcane trash shredding machinery is utilized to incorporate it into the field. The developed trash management machines cut sugarcane leftovers in the field trash of 120 to 150 cm long into small pieces and this practice is environment friendly. A few years ago, sugarcane farmers faced the problem of trash management of the crop. So, they

Figure 2.
Wheat residue handling and management machinery [51, 52].

adopted the conventional method for trash removal from the field and burnt the whole trash in the field. The trash contains very effective nutrients for soil fertility. During trash combustion very hazardous gases such as carbon, nitrogen, and particulate matter (PM) are emitted in the form of smoke which cause injury to human health. The burning of canes trash liberates a significant amount of carbon (CO2) and other greenhouse gases (GHGs). From the burning of cane, it is estimated that direct 10,410 kg/ha carbon emission occurred. Moreover, the 1791 kg CO2/ha is estimated from the other gases (CH4 = 467 kg CO2, CO = 1241 kg, and N2O = 830 kg CO2). The overall carbon emission from the cane burning is summed up to 12,204 kg CO2/ha. This emission is about 37% of the total GHGs of cane production in the farm [56].

Many researchers and industrialists designed and developed new machinery for sugarcane trash handling. Ahmad et al., [57] was designed and developed sugar can trash management machinery to control the environmental pollution and increase the soil fertility. Mukesh & Rani [58] conducted a research study to evaluate the performance evaluation of sugarcane trash management and handling machinery. During the study sugarcane trash shredder cum-chopper machine was investigated for trash management. Nikam [59] developed tractor operated sugarcane trash shredder machinery to manage the field trash. The study concluded that the developed sugarcane trash crusher would be more affordable and suitable for small and medium farmers. Moreover, this machine would be the cheapest option of the sugarcane harvesters now on the market and would result in less sugarcane degradation during storage. Singh et al., [60] worked on the ratoon management machine that was operated with tractor PTO mechanism. That was equipped with stubble-shaving serrated blades mounted on a disc, and two tillage discs for off-barring. The working capacity of the machine was 0.28 ha h^{-1} at the forward speed of 0.67 m/s. **Figure 3** represents the different sugarcane trash management machinery. In order to more effectively reuse waste by shredding and incorporating of sugarcane trash into the soil, particularly in ratoon crops without harming the crop, the tractor-operated two-row rotational sugarcane field shredder by Krishnan & Jayashree, [61]. The developed unit has two rotary members with swinging-type blades to cut the trash.

Figure 3.
Sugarcane residue handling and management machinery [57, 58].

2.4 Cotton stalk management

In Pakistan cotton is the most significant cash crop with high economic impact. In the year 2017–2018, approximately 2699 Mha area cultivated with the cotton cop with annually 11,935 thousand bales production [62]. Due to the large area of cultivation the residue produce in the bulk amount from the field. During the last few years, cotton became non-profitable for farmers due to the attack of pest/disease on cotton crop resulting in low yield and high input cost. The farmers even could not meet their actual expenses [63]. Moreover, mostly in the rural areas not enough energy sources, and cotton sticks are used as burning fuel for cooking and heating purposes. This activity leads to environmental problems such as CO2 and NOx emissions, as well as allowing the pink bollworm to finish its life cycle [64]. Since pulling and gathering the cotton stalks by hand is a time-consuming and laborious process, researchers have worked to create tools and strategies to manage it in the field.

Yumak & Evcim, [65] developed a two row cotton stalk pulling machine and conducted a study for its performance evaluation. The working efficiency of the developed machine was 9.2 ha/hr. by pull the cotton sticks from the field with the 95% efficiency. Gangade et al., [66] conducted a comparative research study of cotton stalk removal tractor operated uprooter, tractor operated slasher, and a tractor-drawn v-blade machine to investigate their working efficiencies for cotton sticks removal from the fied. The outcomes of the study concluded that the working efficiencies of the following machines were 80%, 100%, and 99%, respectively. Sheikh [67] developed a cotton stalk digger that consists of two digging units. The cutting length of the stick digger blade was 0.4 m in length. Ramadan [68] developed a prototype for the for management of the cotton sticks and perform field study to evaluate its working efficiency. The working efficiency with the tilt angle of blade was 45° at the 18.9 m/s rotating speed of under the 19% moisture content. A tractor-drawn cotton stalk

Figure 4.
Cotton stalk residue handling and management machinery [70–72].

puller-cum-chipper was developed by Murugesan et al., [69]. The conduccted study use the rigs test to measure the force requirment for to uproot the cotton stick from the field. Gadir [70] developed a cutting discs base cotton sticks puller machine. The disc for cutting the cotton sticks arrange in the series combination with equal spacing. The implement mounted on the tractor by three pin linkage. According to the findings, the machine worked its maximum (94%) at tilt angles of 30 degrees and rake angles of 20 degrees. It was discovered that 2.8 km/h was the ideal working speed for cotton stalk pulling. Akhtar et al., [71] was designed and developed cotton stalk puller shredder to control the pink boll warm which produces due to cotton stalk trash. The machine tested at three engine speeds and three levels of forward ground speed. A finite element analysis of parts of the cotton stalk puller shredder was performed for the purpose of improving the machine's efficiency. **Figure 4** shows the various cotton stalk handling machinery. Faisal et al., [63] was design and developed the cotton ball stripper to manage the cotton ball from the cotton stalks.

3. Residue management for soil health

Currently, the extensive agriculture attracted much attention and depends on the inputs i.e., machinery, labor. In modern agriculture, farmers use the intensive agriculture to maximize their production by employing the crop rotation. Furthermore, the soil fertility is managed by properly maintain the crop residues. The crop residue management provides the soil surface cover, minimizing soil erosion and protect the soil form the climate effect [8]. Additionally, crop residue manages the physical,

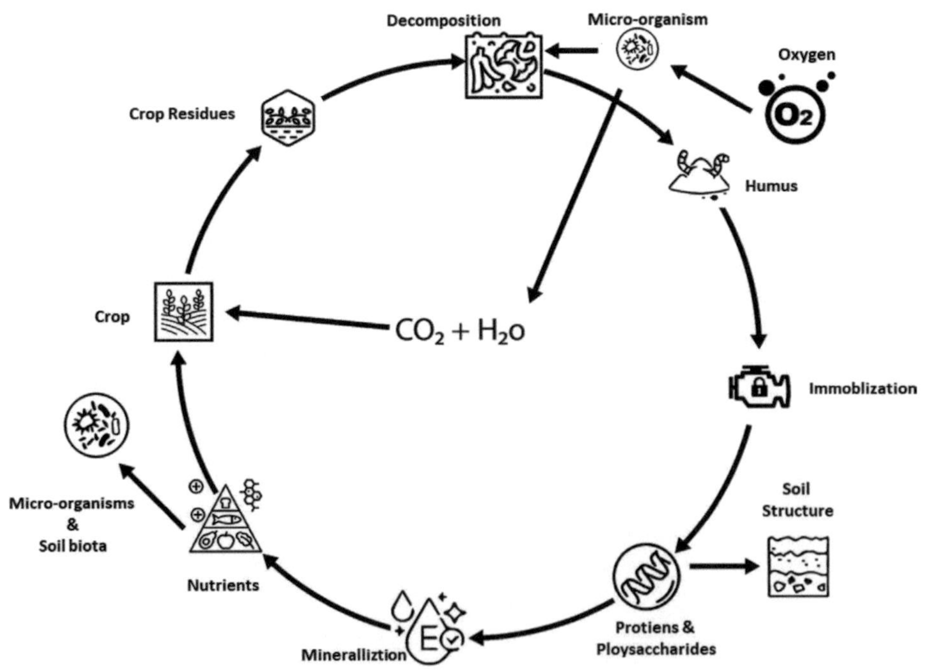

Figure 5.
Illustration of crop residue cycle on soil health profile [76].

biological and chemical process within the soil [73]. If the soil is enriched with the crop residue which ultimately turns into the organic matter boosts the soil fertility and crop production. In contrast, still the crop residue is considered as the waste material due to its less economical value [74]. Crop residues, on the other hand, provide a number of possible pathways for nutrient recycling, including carbon sequestration in soil [8]. The physical, chemical, and biological characteristics of the soil, such as its structure, infiltration rate, plant water availability, nutrient cycling, cation exchange capacity, and species richness of the soil organisms are all improved by the retention of crop residues [22]. Crop residues affect the physical properties of the soil by improving the soil texture, total porosity, and decreasing the bulk density [75]. The chemical properties is also greatly influence the chemical properties i.e., pH, low buffering capacity. [75]. **Figure 5** illustrates the effect of the crop residue cycle on soil condition Pan et al., [75] conducted a one month's incubation experiment to investigate the ameliorating effects on an acidic ultimo with four crop straw decayed products (SDPs), and the results showed that the soil pH increased by 55–75% [77]. Crop residue could potentially increase the organic, carbon, nitrogen, phosphorus, and potassium concentration in the soil. The author conducted the research and concluded that by addition of the straw and partial fertilizers significantly enhances the soil nitrogen up to 20 cm and enhances the of 64% soil fertility cycle [78].

4. Residue management for crop production

Crop residue incorporation is important aspect for the environment friendly agriculture. Moreover, crop residue return can improve crop quality and yields [79]. The meta-analysis was conducted to investigate the effect crop residue mulching on the production quantity [80–82]. Though the number of advantages of the crop residue, but some researcher also investigated that few negative impact of the crop residue

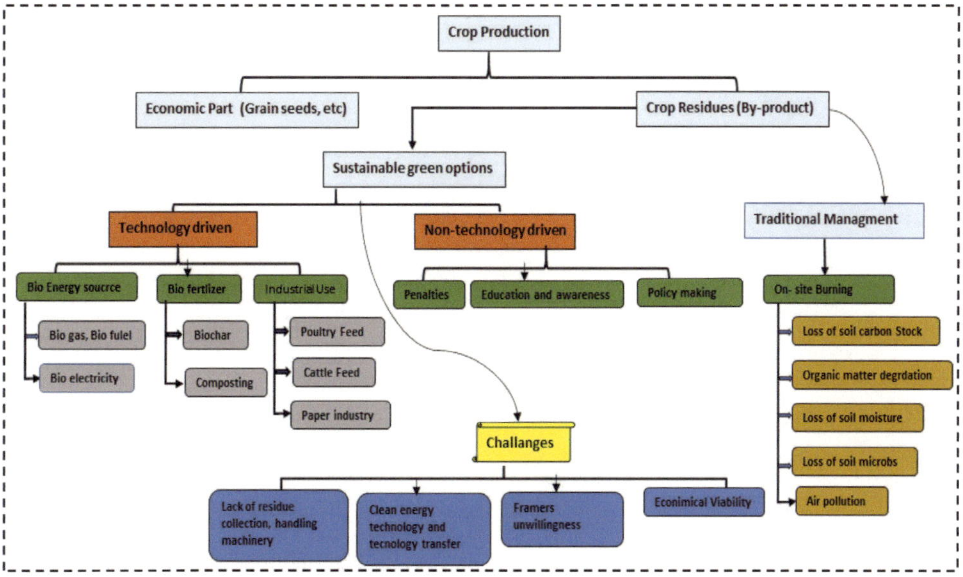

Figure 6.
Framework of crop residue effect on crop production [87].

incorporation on the soil. It disturbs the soil CN ratio [83] which ultimately results in loss of the crop yield [84]. So, in by-product of cereals, straw mulch can enhance a variety of soil texture and crop yield [85].

Aslam et al., [86] studied the impact of different organic mulch and nitrogen sources on wheat crop productivity. As a consequence of these findings, it was concluded that sorghum straw mulch and N in the form of calcium ammonium nitrate can successfully boost wheat productivity in semi-arid locations. The framework of crop residue effect on the crop production is shown in **Figure 6**. Reddy et al., [88] investigated the effect of crop residue on the crop yield enhancement. Mohammad et al., [89] conducted the research study to investigate the effect of tillage and crop residue rotation on the wheat productivity.

5. Residue management for environment protection

The highest portion of greenhouse gas emissions from agriculture is related to emissions from arable soils (5.27%), followed by emissions from animal digestive fermentation (3.21%) and emissions from the use of livestock manure (1.58%) [90]. The adoption of precision farming on a large scale, the cultivation of crops with high carbon sequestration potential, such as energy crops, and grassland management, which offers the potential to store huge amounts of carbon in the soil, and the reforesting of agricultural land are actions that should be taken to reduce greenhouse gas emissions from agricultural practices [91]. Farmers are forced to burn the residue as a result of various socioeconomic, administrative, technical, and commercial issues, which

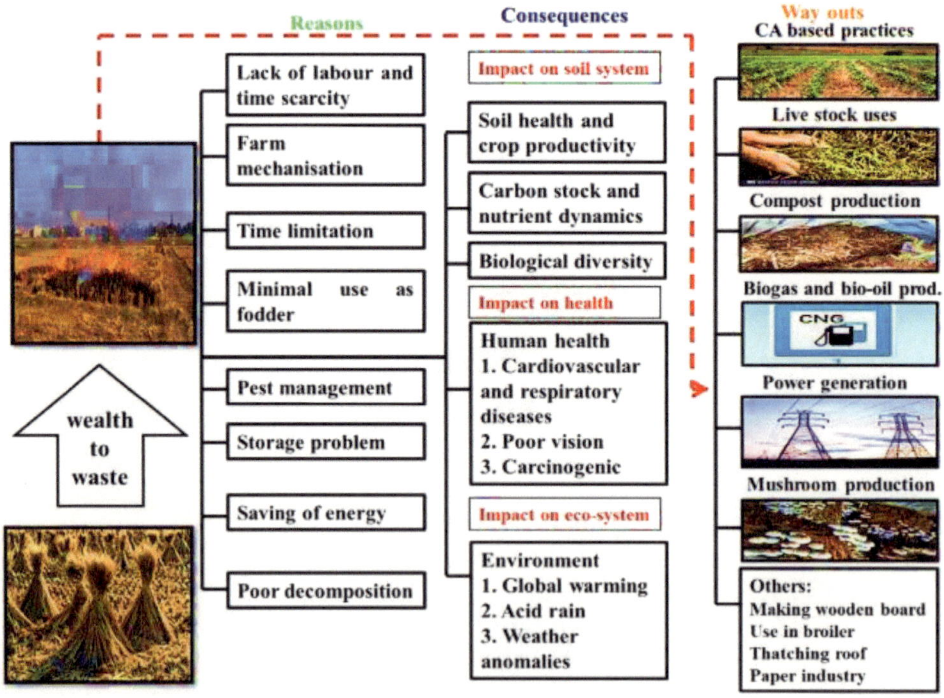

Figure 7.
Residue management protentional for various sectors [92].

causes a number of problems for the ecosystem [92]. In Asia, where rice is preferred over other crops, residue burning is substantially higher than on other continents. This is also true of India, where residue burning is 30% higher than in China (mainland) and 93% higher than in Pakistan [93]. Every single living organism in this area is negatively affected by residual burning. Burning residue has an extremely negative impact on soil health because it causes nitrogen (N) to be converted to nitrate (NO^{-3}) and organic matter to be lost as carbon dioxide (CO2) [92]. The agro-friendly earthworm and other microorganisms in the topsoil are also destroyed when crop residues are burned on site. Moreover, up to 80% and 90%, respectively, of the bacterial and fungal activity is reduced, while up to 65% of the microbial biomass is decreased [94]. **Figure** 7 presents that the crop residue management has the protentional to use conservation agriculture, for live stoke use and energy purpose.

Therefore, academia conducted research studies to control the burning of crop residues to save the greenhouse gas emission and other toxic gases. Malhi et al., [95] studied the effect of tillage and residue management effect on crop yield and greenhouse gas emission. Ashraf et al., [96] studied the carbon emission estimation from rice and wheat cropping. Maucieri et al., [97] provided the meta-analysis on the crop residue management with different tillage systems to control the environmental pollution. The studies find out that the crop residue management in the field have a significant effect on both agriculture and environmental health.

6. Conclusion

Crop residue management and handling in the field is current focus of the conservational agriculture. This study aims to explore solutions for crop residue management and assess the effectiveness of available machinery for handling crop residue. This study has been thoroughly examined the performance of various types of machinery used for collecting and handling agricultural residue which concluded especially, at higher moisture content could be the best condition to chop the crop stubbles after harvesting. The literature of the study concluded that the use of machinery for residue management saves the labore requirement and timely sow the next crop that enhance the crop production. Moreover, the residue incorporation into the soil improves the soil health, reduce greenhouse gas emissions, and save environmental pollution. In addition, the performance of these machines depends on various factors such as fuel consumption, chopping capacity, and moisture content of soil and trash. This study provides the guidelines to the researchers and industrial stallholders should be work on small-scale trash management machinery.

Author details

Fiaz Ahmad[1†*], Aftab Khaliq[1†], Ding Qishuo[2] and Muhammad Sultan[1]

1 Department of Agricultural Engineering, Bahauddin Zakariya University, Multan, Pakistan

2 College of Engineering, Nanjing Agricultural University, Nanjing, China

*Address all correspondence to: fiazahmad@bzu.edu.pk

† These authors contributed equally to this work.

References

[1] Shinde R, Shahi DK, Mahapatra P, Singh CS, Naik SK, Thombare N, et al. Management of Crop Residues with special reference to the on-farm utilization methods: A review. Industrial Crops and Products. 2022;**181**:114772. DOI: 10.1016/j.indcrop.2022.114772

[2] Rafiq M, Ahmad F, Atiq M. The determinants of the crop residue Management in Pakistan: An environmental appraisal. Business and Economic Review. 2019;**11**:179-200. DOI: 10.22547/ber/11.4.8

[3] Johnson JMF, Novak JM, Varvel GE, Stott DE, Osborne SL, Karlen DL, et al. Crop residue mass needed to maintain soil organic carbon levels: Can it Be determined? Bioenergy Research. 2014;**7**:481-490

[4] Lal R. World crop residues production and implications of its use as a biofuel. Environment International. 2005;**31**:575-584

[5] Haddaway NR, Hedlund K, Jackson LE, Kätterer T, Lugato E, Thomsen IK, et al. How does tillage intensity affect soil organic carbon? A systematic review. Environmental Evidence. 2017;**6**(30):1-48. DOI: 10.1186/s13750-017-0108-9

[6] Bhattacharyya P, Barman D. Crop Residue Management and Greenhouse Gases Emissions in Tropical Rice Lands. Amsterdam, Netherlands: Elsevier Inc.; 2018; ISBN 9780128121290

[7] Snyder CS, Bruulsema TW, Jensen TL, Fixen PE. Review of greenhouse gas emissions from crop production systems and fertilizer management effects. Agriculture, Ecosystems and Environment. 2009;**133**:247-266

[8] Sarkar S, Skalicky M, Hossain A, Brestic M, Saha S, Garai S, et al. Management of crop residues for improving input use efficiency and agricultural sustainability. Sustainability. 2020;**12**:1-24. DOI: 10.3390/su12239808

[9] Zafar S. Rice Straw as Bioenergy Resource. BioEnergy Consult. Retrieved from World Wide Web https://www.bioenergyconsult.com/rice-straw-as-bioenergy-resource_2018.India

[10] Singh Y, Sidhu HS. Management of Cereal Crop Residues for sustainable Rice-wheat production system in the indo-Gangetic Plains of India. Proceedings of the Indian National Science Academy. 2014;**80**:95-114

[11] Cavalett O, Norem Slettmo S, Cherubini F. Energy and environmental aspects of using eucalyptus from Brazil for energy and transportation Services in Europe. Sustainability. 2018;**10**:4068

[12] Garg IK. Design and development of Rice straw chopper-cum-spreader. Journal of Research. 2004;**41**:130-138

[13] Jia H, Wang L, Li C, Tan H, Ma C. Combined stalk–stubble breaking and mulching machine. Soil and Tillage Research. 2010;**107**:42-48

[14] Wang H. Study on 2BMFJ-3 Type No-Till Soybean Precision Planter with Straw-Covering in Wheat Stubble Fields. PhD diss. Heilongjiang, China: Northeast Agric. Univ. Dep. Eng; 2013

[15] Elfatih A, Arif EM, Atef AE. Evaluate the modified chopper for Rice straw composting. Journal of Applied Sciences Research. 2010;**6**:1125-1131

[16] Verma A, Singh A, Singh A, Sidhu GS, Dixit A. Performance

evaluation of tractor operated Paddy straw Mulcher. Journal of Krishi Vigyan. 2016;**4**:70-75

[17] Wegner BR, Chalise KS, Singh S, Lai L, Abagandura GO, Kumar S, et al. Response of soil surface greenhouse gas fluxes to crop residue removal and cover crops under a corn–soybean rotation. Journal of Environmental Quality. 2018;**47**:1146-1154

[18] Lehtinen T, Schlatter N, Baumgarten A, Bechini L, Krüger J, Grignani C, et al. Effect of crop residue incorporation on soil organic carbon and greenhouse gas emissions in European agricultural soils. Soil Use and Management. 2014;**30**:524-538

[19] Zhang C, Zeng G, Huang D, Lai C, Chen M, Cheng M, et al. Biochar for environmental management: Mitigating greenhouse gas emissions, contaminant treatment, and potential negative impacts. Chemical Engineering Journal. 2019;**373**:902-922

[20] Turmel M-S, Speratti A, Baudron F, Verhulst N, Govaerts B. Crop residue management and soil health: A systems analysis. Agricultural Systems. 2015;**134**:6-16

[21] Zahid A, Ali S, Ahmed M, Iqbal N. Improvement of soil health through residue management and conservation tillage in Rice-wheat Cropping system of Punjab, Pakistan. Agronomy. 1844;**2020**:10

[22] Bisen N, Rahangdale CP. Crop residues management option for sustainable soil health in Rice-wheat system: A review. International Journal of Chemical Studies. 2017;**5**:1038-1042

[23] Clay DE, Alverson R, Johnson JMF, Karlen DL, Clay S, Wang MQ, et al. Crop residue management challenges: A special issue overview. Agronomy Journal. 2019;**111**:1-3

[24] Yadav AS, Arora S. Crop residue Management in Diverse Agro-Ecosystems for improving soil health-an overview. Journal of Soil and Water Conservation. 2018;**17**:387-392

[25] Paul BK, Vanlauwe B, Ayuke F, Gassner A, Hoogmoed M, Hurisso TT, et al. Medium-term impact of tillage and residue management on soil aggregate stability, soil carbon and crop productivity. Agriculture, Ecosystems and Environment. 2013;**164**:14-22

[26] Piccoli I, Sartori F, Polese R, Berti A. Crop yield after 5 decades of contrasting residue management. Nutrient Cycling Agroecosystems. 2020;**117**:231-241

[27] Hiel M-P, Chélin M, Parvin N, Barbieux S, Lemtiri A, Colinet G, et al. Crop residue Management in Arable Cropping Systems under a temperate climate. Part 2: Soil physical properties and crop production. A review. Biotechnologie, Agronomie, Société et Environnement. 2016;**20**(SI):246-256. DOI: 10.25518/1780-4507.12986

[28] Koga N, Tsuji H. Effects of reduced tillage, crop residue management and manure application practices on crop yields and soil carbon sequestration on an Andisol in northern Japan. Soil Science & Plant Nutrition. 2009;**55**:546-557

[29] Devi S, Gupta C, Jat SL, Parmar MS. Crop residue recycling for economic and environmental sustainability: The case of India. Open Agriculture. 2017;**2**:486-494

[30] Fazio S, Barbanti L. Energy and economic assessments of bio-energy systems based on annual and perennial crops for temperate and tropical areas. Renewable Energy. 2014;**69**:233-241

[31] Chauhan A, Upadhyay S, Saini G, Senthilkumar N. Agricultural crop residue based biomass in India: Potential assessment, methodology and key issues. Sustainable Energy Technologies and Assessments. 2022;**53**:102552

[32] Ahiduzzaman M, Islam AKM, Yaakob Z, Ghani JA, Anuar N. Agricultural residues from crop harvesting and processing: A renewable source of bio-energy. Biomass and Bioenergy. Switzerland: Springer; 2014. pp. 323-337

[33] Shan B-S, Johnson-Beebout YH, Yadvinder-Singh SE, Buresh RJ. Crop residue Management for Lowland Rice-Based Cropping Systems in Asia. Advances in Agronomy. 2008;**98**:117-199. DOI: 10.1016/S0065-2113(08)00203-4

[34] Cropping P. Crop residue Management for Nutrient Cycling and Improving Soil Productivity in Rice-based Cropping Systems in the Tropics. Advances in Agronomy. 2005;**85**:85005-85006

[35] Roy P, Kaur M. Status and problems of Paddy straw Management in West Bengal. International Journal of Advances in Agricultural and Environmental Engineering. 2015;**2**:144-152

[36] Chaudhary A, Chhokar RK, Yadav DB, Sindhu VK, Ram H, Rawal S, et al. In-situ Paddy straw management practices for higher resource use efficiency and crop productivity in indo-Gangetic Plains (IGP) of India. Wheat Barley Research. 2019;**11**:172-198. DOI: 10.25174/2249-4065/2019/96323

[37] Hegazy RA, Dhaliwal IS. Evaluation of a power driven residue manager for No-till drills. Agricultural Engineering International: CIGR Journal. 2011;**13**(1):1641,1-1641,7

[38] Sidhu HS, Manpreet-Singh, Humphreys E, Yadvinder-Singh, Balwinder-Singh, Dhillon SS, et al. Malkeet-Singh; Sarbjeet-Singh the happy Seeder enables direct Drilling of Wheat int,o Rice stubble. Australian Journal of Experimental Agriculture. 2007;**47**:844-854. DOI: 10.1071/EA06225

[39] Regatti Venkat SSM. Happy Seeder: A Residue. Management Technology. 2019;**8**:1959-1962

[40] Kathpalia J, Tyagi R, Saroha AK. Crop residue management with reference to happy Seeder in Kaithal District: A sociological study. The Pharma Innovation. 2021;**10**:736-744

[41] Warnatzsch EA, Reay DS. Assessing climate change projections and impacts on Central Malawi's maize yield: The risk of maladaptation. Science of the Total Environment. 2020;**711**:134845

[42] Scarlat N, Dallemand J-F, Monforti-Ferrario F, Nita V. The role of biomass and bioenergy in a future bioeconomy: Policies and facts. Environment and Development. 2015;**15**:3-34

[43] Rasse DP, Rumpel C, Dignac M-F. Is soil carbon mostly root carbon? Mechanisms for a specific stabilisation. Plant and Soil. 2005;**269**:341-356

[44] Santhi C, Srinivasan R, Arnold JG, Williams JR. A Modeling approach to evaluate the impacts of water quality management plans implemented in a watershed in Texas. Environmental Modelling & Software. 2006;**21**:1141-1157

[45] Panagos P, Borrelli P, Meusburger K, Alewell C, Lugato E, Montanarella L. Estimating the soil erosion cover-management factor at the European scale. Land Use Policy. 2015;**48**:38-50

[46] Unger JS, Glasner C. Cost analysis of chaff harvesting concepts in Germany. Agronomy. 2019;**9**:579

[47] Pari L, Alfano V, Scarfone A, Bergonzoli S, Suardi A, Lazar S. Best available technologies to harvest cereal chaff. In: Proceedings of the 26th European Biomass Conference and Exhibition (EUBCE). Copenhagen, Denmark. Florence, Italy: ETA-Florence Renewable Energies; 2018. pp. 14-18

[48] Muth DJ Jr, Bryden KM. An integrated model for assessment of sustainable agricultural residue removal limits for bioenergy systems. Environmental Modelling & Software. 2013;**39**:50-69

[49] Suardi A, Bergonzoli S, Alfano V, Scarfone A, Pari L. Economic distance to gather agricultural residues from the field to the integrated biomass logistic Centre: A Spanish case-study. Energies. 2019;**12**:3086

[50] Zhang YX, Zeng XY, Yang XZ, Gao ZY, Li BQ. Experimental study on straw chopping and tillage stubble machine [in Chinese with English abstract]. Journal of Henan Agricultural University. 2002;**36**(2):179-182

[51] Mahmood HS, Ahmad T, Ali Z, Ahmad M, Amjad N. Field evaluation of a wheat straw chopper. Pakistan Journal of Agricultural. 2016;**29**:298-312

[52] Suardi A, Stefanoni W, Bergonzoli S, Latterini F, Jonsson N, Pari L. Comparison between two strategies for the collection of wheat residue after mechanical harvesting: Performance and cost analysis. Sustainability. 2020;**12**(4936):1-17. DOI: 10.3390/SU12124936

[53] Qureshi MA, Afghan S. Sugarcane cultivation in Pakistan. In: Sugar B, editor. Pakistan Sugar Book. Karachi, Pakistan: Pakistan Society of Sugar Technologists; 2005

[54] Mohan P, Ponnusamy D. Addressing the challenges of sugarcane trash decomposition through effective microbes. In: Proceedings of the Proceedings of the 2011 International Conference on Food Engineering and Biotechnology. Singapore: IACSIT Press; 2011. pp. 229-233

[55] Cançado JED, Saldiva PHN, Pereira LAA, Lara LBLS, Artaxo P, Martinelli LA, et al. The impact of Sugar cane-burning emissions on the respiratory system of children and the elderly. Environmental Health Perspectives. 2006;**114**:725-729. DOI: 10.1289/ehp.8485

[56] Arsenio FM, Acuna A, City B, Occidental N, Granja L, City Q. Detrashing sugarcane stalks: The 1 St critical step to the shift to organic farming in sugarcane. Production. 2016;**12**:1057-1070

[57] Ahmad et al. In-field sugarcane trash management strategies to control environmental pollution: A short review. In: Proceedings of the International Conference on Energy, Water and Environment – ICEWE-2021. Lahore: Dept., mechanical engineering UET; 2021. pp. 375-376

[58] Mukesh S, Rani V. Performance evaluation of machinery for sugarcane handling and trash management. International Agricultural Engineering Journal. 2017;**10**:234-238. DOI: 10.15740/has/ijae/10.1/234-238

[59] Nikam G. Design and fabrication of sugarcane trash removal machine. International Journal of Engineering Research & Technology (IJERT). 2021;**10**(9):19-25

[60] Singh AK, Singh PR, Solomon S. Design and development of a tractor-operated disc-type sugarcane ratoon management device. Sugar Tech. 2017;**19**:501-509. DOI: 10.1007/s12355-016-0483-5

[61] Krishnan DA, Jayashree GC. Design and development of tractor mounted sugarcane trash shredder. The Madras Agricultural Journal. 2012;**99**:901-904

[62] Ahmad F, Jamil MT, Khan AA, Mehmood-Khan Z, Qiu B, Ma J, et al. Field performance evaluation and finite element simulation of cotton stalk puller-shredder: A sustainable mechanical solution to control pink bollworm (Pectinophora Gossypiella). Sustainability. 2020;**12**(3407):1-25. DOI: 10.3390/SU12083407

[63] Faisal MAH et al. Study on performance evaluation of cotton boll stripper. In: Proceedings of the 6th International Conference on Energy, Environment and Sustainable Development 2022 (EESD 2022) Study. Vol. 2022. Jamshoro, Pakistan: Mehran University of Engineering and Technology; 2022. pp. 223-227

[64] Guoliang CAO, Zhang X, Sunling G, Zheng F. Investigation on emission factors of particulate matter and gaseous pollutants from crop residue burning. Journal of Environmental Sciences. 2008;**20**:50-55

[65] Yumak H, Evcim Ü. A two-row cotton stalk pulling machine. In: Proceedings of the International Congress on Mechanization and Energy in Agriculture. Adana, Turkey, 1-4 October 1990. Turkey: Ministry of Agriculture, Forestry and Rural Affairs; 1990. pp. 416-425

[66] Gangade CN, Patil PG, Dhanphule SS. Comparative study of different methods of cotton stalk removing. PKV Research Journal. 2000;**24**:87-89

[67] Sheikh EDAG. El development and performance of 2-unit diggers for cotton stalks uprooting and groundnut lifting. AMA, Agricultural Mechanization in Asia, Africa and Latin America. 2000;**31**:46-52

[68] Ramadan YYR. Development and evaluation of a cotton stalks puller. Journal of Soil Sciences and Agricultural Engineering. 2010;**1**:1061-1073

[69] Murugesan R, Shukla SK, Patil PG, Arude VG. Design and development of tractor-drawn cotton stalk puller-cum-chipper. Agriculture Engineering Today. 2004;**28**:1-6

[70] Gadir E. Development of tractor operated cotton stalk puller. American Journal of Experimental Agriculture. 2013;**3**:495-505. DOI: 10.9734/ajea/2013/2568

[71] Akhtar A, Sultan U, Rudra RP, Ehsan F, Kashif M, Mohsin M, et al. Structural analysis of cotton stalk puller and shredder machine. Alexandria Engineering Journal. 2023;**64**:335-347. DOI: 10.1016/j.aej.2022.09.002

[72] Khedkar PA, Khedkar SS. Design and fabrication of cotton stalks cutting and crushing tool for power tiller. Journal of Research Engineering and Applied Science. 2019;**04**:122-126. DOI: 10.46565/jreas.2019.v04i03.006

[73] Purwanto BH, Alam S. Impact of intensive agricultural management on carbon and nitrogen dynamics in the humid tropics. Soil Science & Plant Nutrition. 2020;**66**:50-59

[74] Kumar K, Goh KM. Crop residues and management practices: Effects on soil quality, soil nitrogen dynamics, crop

yield, and nitrogen recovery. Advances in Agronomy. 1999;**68**:197-319

[75] Fu B, Chen L, Huang H, Qu P, Wei Z. Impacts of crop residues on soil health: A review. Environmental Pollutants and Bioavailability. 2021;**33**:164-173. DOI: 10.1080/26395940.2021.1948354

[76] Lavelle P, Dangerfield M, Fragoso C, Eschenbrenner V, Lopez-Hernandez D, Pashanasi B, et al. The relationship between soil macrofauna and tropical soil fertility. In: Woomer PL, Swift MJ, editors. The Biological Management of Tropical Soil Fertility. Chichester, UK: John Wiley & Sons; 1994. pp. 137-169

[77] Pan X-Y, Shi R-Y, Hong Z-N, Jiang J, He X, Xu R-K, et al. Characteristics of crop straw-decayed products and their ameliorating effects on an acidic Ultisol. Archives of Agronomy and Soil Science. 2021;**67**(12):1708-1721

[78] Zhao X, Yuan G, Wang H, Lu D, Chen X, Zhou J. Effects of full straw incorporation on soil fertility and crop yield in Rice-wheat rotation for silty Clay loamy cropland. Agronomy. 2019;**9**:133

[79] Xu J, Han H, Ning T, Li Z, Lal R. Long-term effects of tillage and straw management on soil organic carbon, crop yield, and yield stability in a wheat-maize system. Field Crops Research. 2019;**233**:33-40

[80] Li Q, Li H, Zhang L, Zhang S, Chen Y. Mulching improves yield and water-use efficiency of potato Cropping in China: A meta-analysis. Field Crops Research. 2018;**221**:50-60

[81] Yu Y-Y, Turner NC, Gong Y-H, Li F-M, Fang C, Ge L-J, et al. Benefits and limitations to straw-and plastic-film mulch on maize yield and water use efficiency: A meta-analysis across hydrothermal gradients. European Journal of Agronomy. 2018;**99**:138-147

[82] Zhao X, Liu B, Liu S, Qi J, Wang X, Pu C, et al. Sustaining crop production in China's cropland by crop residue retention: A meta-analysis. Land Degradation & Development. 2020;**31**:694-709

[83] Lu X. A meta-analysis of the effects of crop residue return on crop yields and water use efficiency. PLoS One. 2020;**15**:1-18. DOI: 10.1371/journal.pone.0231740

[84] Mohsin AU, Ahmad J, Ahmad AUH, Ikram RM, Mubeen K. Effect of nitrogen application through different combinations of urea and farm yard manure on the performance of spring maize (Zea Mays L.). Journal of Animal and Plant Sciences. 2012;**22**:195-198

[85] Kosterna E. Organic mulches in the vegetable cultivation (a review). Ecological Chemistry and Engineering A. 2014;**21**(4):481-492

[86] Aslam MU, Chattha MU, Khan I, Maqbool R, Chattha MB, Hussan F, et al. Effect of different organic mulches and nitrogen sources on the productivity of wheat crop grown in semi-arid area. Pakistan Journal of Agricultural Research. 2021;**34**:648-655. DOI: 10.17582/journal.pjar/2021/34.3.648.655

[87] Sarkar S, Skalicky M, Hossain A, Brestic M, Saha S, Garai S, et al. Management of Crop Residues for Improving Input Use Efficiency and Agricultural Sustainability. 2020;**12**:1-24. DOI: 10.3390/su12239808

[88] Reddy BVS, Reddy PS, Bidinger F, Blümmel M. Crop management factors influencing yield and quality of crop residues. Filed Crops Research, Elsevier. 2003;**84**:57-77

[89] Mohammad W, Shah SM, Shehzadi S, Shah SA. Effect of tillage, rotation and crop residues on wheat crop productivity, fertilizer nitrogen and water use efficiency and soil organic carbon status in dry area (Rainfed) of north-West Pakistan. Journal of Soil Science and Plant Nutrition. 2012;**12**:715-727

[90] Rutkowska B, Szulc W, Sosulski T, Skowrońska M, Szczepaniak J. Impact of reduced tillage on CO2 emission from soil under maize cultivation. Soil and Tillage Research. 2018;**180**:21-28. DOI: 10.1016/j.still.2018.02.012

[91] Frank S, Havlík P, Soussana J-F, Levesque A, Valin H, Wollenberg E, et al. Reducing greenhouse gas emissions in agriculture without compromising food security? Environmental Research Letters. 2017;**12**:105004

[92] Dutta A, Patra A, Hazra KK, Nath CP, Kumar N, Rakshit A. A state of the art review in crop residue burning in India: Previous knowledge, present circumstances and future strategies. Environmental Challenges. 2022;**8**:100581. DOI: 10.1016/j.envc.2022.100581

[93] Flammini A, Pan X, Tubiello FN, Qiu SY, Rocha Souza L, Quadrelli R, et al. Emissions of greenhouse gases from energy use in agriculture, forestry and fisheries: 1970-2019. Earth System Science Data. 2022;**14**:811-821

[94] Anand A, Kumar V, Kaushal P. Biochar and its twin benefits: Crop residue management and climate change mitigation in India. Renewable and Sustainable Energy Reviews. 2022;**156**:111959. DOI: 10.1016/j.rser.2021.111959

[95] Malhi SS, Lemke R, Wang ZH, Chhabra BS. Tillage, nitrogen and crop residue effects on crop yield, nutrient uptake, soil quality, and greenhouse gas emissions. Soil and Tillage Research. 2006;**90**:171-183

[96] Ashraf MN, Mahmood MH, Sultan M, Shamshiri RR, Ibrahim SM. Investigation of energy consumption and associated CO 2 emissions for wheat – Rice crop rotation farming. Energies. 2021;**14**:1-18. DOI: 10.3390/en14165094

[97] Maucieri C, Tolomio M, McDaniel MD, Zhang Y, Robatjazi J, Borin M. No-tillage effects on soil CH4 fluxes: A meta-analysis. Soil and Tillage Research. 2021;**212**:105042. DOI: 10.1016/j.still.2021.105042

Chapter 2

Food Wastage Footprint, Food Security, Environment and Economic Growth Nexus in Developing Countries

Muhammad Ramzan Sheikh, Neelam Asghar Ali and Asma Aslam

Abstract

FAO estimates that each year, approximately one-third of all food produced for human consumption in the world is lost or wasted. This food wastage represents a missed opportunity to improve global food security, but also to mitigate environmental impacts and resource use from food chains. This chapter attempts to identify the impact of food wastage, food security, and environmental quality on economic growth of developing countries from 1990 to 2021. Method of Moments Quantile Regression (MMQR) has been used to estimate the results. The findings indicate that food wastage, food access, and environment have a negative effect while food availability and food security have a positive effect on economic growth. To the best of our knowledge, no study has yet analyzed the impacts of food wastage on economic growth from an environmental and food security perspective. This study provides a worldwide account of the environmental footprint of food wastage along the food security, focusing on impacts on environmental quality, as well as an economic quantification based on economic growth.

Keywords: GDP per capita, food wastage, poverty headcount ratio, prevalence of undernourishment, environmental quality

1. Introduction

Generally, food waste is considered as all the foodstuffs that have been discarded from the chain of food supply but still are fit for human consumption and perfectly edible. These goods are eventually eliminated and disposed of for some esthetic or economic reasons or may be due to the closeness to the "use by" or "best before" date. As a result, it creates negative externalities from the perspective of the environment and can raise economic costs along with revenue loss for the industries [1].

Food waste may occur at any stage such as processing, handling, packing, storing, or transportation. Many people are involved in different stages of production and consumption such as farmers, distributors, consumers, retailers, and wholesalers. Therefore, all these entities can be responsible for reducing the wastage of food due to their negligence or any other reason(s). Every year billions of tons of food usually are wasted globally [2]. In 2014, approximately, 6.68 billion tons of food were wasted; in 2015, 6.72 billion tons; and 6.92, 7.01, and 6.86 billion tons were wasted from 2016 to 2018 respectively [3]. The category of food wastage includes cereals, fruits, vegetables, spices, oilseeds, pulses, milk, roots as well as meat products [4].

Food waste asserts a negative effect on food security [5, 6]. In developing nations, a significant quantity of food is usually discarded by industries and households, and at a local level, it has been observed that better management and reduction of food wastage can lessen the food security of businesses and households [7–10]. Food waste can reduce the availability of food and therefore, have an adverse effect on the environment, consumption chain, and the resources used for the production of food. About one-quarter of water, land, and fertilizer for the production of the crop, even though environmental and resource constraints are likely to limit the production of food universally [11]. In fact, at all stages of the food supply chain, the impact of food waste on the environment can occur, but the extent of its effect at different stages of the supply chain can vary according to the economic development and environmental dimensions of a country. In industrialized nations, most of the food is wasted at the end of the supply chain process, and it brings large food waste and environmental degradation. In developing nations, the reduction measures that target farm losses can be most operative in decreasing the environmental tracks of food waste. The trade-off between environmental sustainability and food security is likely to be intensified in future because of many universal transactions, dietary changes, wealthier and large populations, rising interdependence of energy and food as well as the competition between climate change, food, and bioenergy over the resources. Ensuring the accessibility of safe and high-quality food around the globe, specifically in face of climate change, needs adequate facilities of the cold chain [12]. According to an estimation by the International Institute of Refrigeration 2009, if developing nations attained the same capacities of cold chain same as developed nations, approximately 200 million tons of food can be saved annually.

2. Concepts and definitions

The concept and definitions of various terms are as follows:

2.1 Food waste

What is food waste? How can we define it? Although there are many definitions in the literature, no common definition of food waste exists. The existing definitions vary in different aspects such as: Which type of food would be considered? Only edible parts would be included? Food diverted to other usage can be considered food waste? How the food loss differs from food wastage?

Food wastage is a general term that encapsulates both food loss and food waste. Although the difference between food loss and food waste is relevant conceptually, it is beneficial from a policy point of view.[1] Food waste is a result of purchasing decisions by the consumers, or the decisions by food service providers and retailers that can affect the behavior of consumers. Various definitions of food waste can be stated as:

Surplus food that cannot be used for feeding humans or animals, or surplus food that would not be recovered or recycled in any form and is usually disposed of [13].

Food waste is a reduction in the quality or quantity of food that results from the actions as well as decisions by the consumers, retailers, or food service providers. Quantitative food waste is a physical reduction in the masses of food that results from the decisions taken by consumers, retailers, and food services while qualitative food waste is a decrease in food characteristics that reduced the food value in terms of its use.

2.2 Food security

Food security exists at all times, when all people, have economic, social, and physical access to safe, sufficient, and nutritious food to meet their nutritional requirements and preferences of food for a healthy and active life. This concept pinpoints four basic aspects of food security, i.e., food stability, food access, food availability, and food utilization [14, 15].

2.3 Availability

Availability reports whether or not the food is potentially, actually, or physically present and contains the features of food reserves, production, transportation, markets, and wild foods.

2.4 Access

If food is potentially, actually, or physically present, the next question is whether individuals or households have access to the food.

2.5 Utilization

The next issue after assuring that food is available and the households have suitable access to it, is whether or not the households are maximizing their intakes of sufficient energy and nutrition. Sufficient nutrients and energy by the individuals are a result of better feeding, care, dietary diversity as well as intra-household food distributions.

2.6 Stability

If all the above conditions are met, then stability is a situation in which the entire system is steady, hence ensuring that at all times households are food secure. Economic, social, climatic, and political factors can be the sources of instability.

[1] Food loss is actually a result of actions and decisions of the suppliers that influences the food supply; if the food loss decreases, the food supply rises. Food waste is linked with consumers or consumption.

2.7 Environmental sustainability

Environmental sustainability is associated with protecting the integrity of the ecosystem along with confirming that the activities of society remain within the carrying capacity of the ecological system of the earth [16–18]. In this concept, key metrics involved freshwater use, change in climate, change in land use, and the flows of nutrients via the ecosystem. The universal system of food uses a huge amount of resources to satisfy the demand of consumers for food, and the major contributors that influence the environment are biodiversity loss, phosphorus cycles, climate change, and disrupted nitrogen [19].

3. Review of assorted studies

This section reviews the various studies on food waste, economic growth, food security, and environmental quality.

3.1 Studies based on food waste and economic growth

In the twenty-first century, the management of food waste is one of the universal challenges due to its antagonistic effects on the economy and environment. A study investigated the effect of food wastage on the economic growth of 165 nations by using OLS and GLM techniques and found out the negative effect of food wastage as well as poverty on GDP growth [4]. Similarly, in Japan, it was observed that technology has needed in dropping and reutilizing food waste while studying the interrelationship between sustainable community and food waste [20]. A study done on Indonesia over the period 2000–2019 indicated that the combination of food loss reduction and yield improvement can raise the GDP of Indonesia by 0.2% approximately. In a comparison of increasing productivity and decreasing food loss, they prefer the reduction of food loss for increasing GDP [21]. Contrary to these studies, the social impact along with the economic impact of food wastage in the United States and South Korea was analyzed and concluded that in the United States, if a national strategy would not be adopted to remedy the waste, the food waste endures being a major issue [22]. However, South Korea embarked on an essential paradigm turning food wastage into an influential economic driver and leading to a fast decile in the waste of food [23]. Another study was carried on Sri Lanka's hotel industry to examine the influence of food waste on the operations of sustainable business in 2009. This study indicated that food waste is common in the industry of hospitality since it usually has a great focus on the satisfaction of customers through the provision of high-quality foods, which in turn can depress their businesses [9].

3.2 Studies based on food waste and food security

Food security is a universal challenge. Current investments in genetic engineering and technologies of agriculture, which frequently focus on increasing productivity through crop intensification, expanded cropland, and higher yields, have contributed to the enhancement of crop efficiency and will contribute to meet the challenge of food production. Food wastage has a negative and significant influence on food security by considering urban women's role in the management of households in Tehran [24]. Two factors that influence food waste are the number of family members and the education of farmers. So, it has been suggested that government would focus on farmers' requirements to support food security [25]. Additionally, the significance of food waste and food loss inhibition as complementary solutions have been reported as a big challenge of environmental sustainability and food security and concluded that the reduction of food loss and food waste holds excessive potential for boosting food security [26]. Along with food security, the effect of reducing food loss and waste on greenhouse gas emissions, trade, and land use is also studied. Authors have concluded that there is substantial potential scope for reducing food waste and food loss on food security as well as greenhouse gas emissions [27]. On the other hand, it was also observed that lessening food waste could not only decrease the depletion and degradation burden of natural resources but can also raise food security by increasing the production of food by 60% [28].

3.3 Studies based on food wastage and environment

The environmental effect of food consumption and production is usually aggravated when food is wasted instead of consumed, and the environmental effect of food wastage covers all the emissions that have derived from various steps of the food supply chain. Many studies have evaluated the effect of food waste on the environment in different aspects. The effects on the environment due to food waste reduction were observed by using a systematic dynamic approach, and it was concluded that through a 1% reduction in the scenario of food waste, we can control the pollution emissions by 0.82% and save the energy demand by 0.96% [29]. Some studies also support the previous piece of literature that concluded that animal-based food wastage has a great environmental footprint in contrast to food wastage based on plants. This was analyzed during their assessment of the environmental influence of food wastage in Turkey [30]. The impact of global food waste and food surplus to tackle environmental and economic sustainability on seven nations, six organizations, and six continents was observed, which provides a framework to identify the crisis of environmental and economic sustainability. The authors find out that the selected organizations and regions were relevant to the fundamental problems of food waste, food surplus economic sustainability, and environmental sustainability. Further, through the globalization trend, this study pinpointed that the reduction of food wastage may help out for regulating environmental sustainability [7]. Alternative to these empirical studies, another study probed the environmental effects of technologies for food waste treatment in the United States. Findings revealed that overall burdens on the environment may be decreased through the source of separating food waste technologies [31]. Moreover, the authors recognized that climate change, resource depletion, environmental acidification and eutrophication, and ozone depletion are also influenced by food waste. Additionally, the environmental influence of food wastage in Europe was also analyzed. Results have confirmed that most of the environmental effects have been derived from the initial production step of the chain. Approximately three-quarters of food wasted is related to effects of the global warming initiates by greenhouse emissions during the steps of production. Emissions by the processing of food activities contribute 6%, distribution and retail 7%, consumption of food 8%, and food disposal 6% to the food wastage associated impacts [32]. It has also been exposed that the wastewater used for farming as well as irrigation of adjacent areas and the water drained from the kitchens of the restaurants were polluted considerably and were not suitable for irrigation, aquatic organisms, and agricultural purposes in India. This study has been carried out on five sampling points of canteens, restaurants, confectionaries, and famous hotels from January 2017 to December 2017 [33].

4. Model, data, and methodology

The general quantile conditional function for quantile τ is given as:

$$Q_{GDPG_{it}}(\tau|\gamma_i, \delta_t, X_{i,t}) = \gamma_i + \delta_t + \alpha_{1,\tau}LAB_{i,t} + \alpha_{2,\tau}CAP_{it} + \alpha_{3,\tau}SSE_{i,t} + \alpha_{4,\tau}FOOD_{W_{i,t}} + \alpha_{5,\tau}FOOD_{AC_{i,t}} + \alpha_{6,\tau}FOOD_{AV_{i,t}} + \alpha_{7,\tau}FOOD_{ST_{i,t}} + \alpha_{8,\tau}FOOD_{UT_{i,t}} + \alpha_{9,\tau}CO2_{i,t} + \mu_{\tau,i,t} \quad (1)$$

where τ show quantiles such as 25th, 50th 75th and 90th $i = 1, \dots \dots \dots N$ is for cross-sections, and t for the time-period starting from $t = 1, \dots \dots T$, $GDPG_{it}$ is the dependent variable. Equations for each quantile are given as:

$$Q_{0.25}(GDPG_{it}) = \gamma_{0.25} + \alpha_{1,0.25}LAB_{i,t} + \alpha_{2,0.25}CAP_{it} + \alpha_{3,0.25}SSE_{i,t} + \alpha_{4,0.25}FOOD_{W_{i,t}} + \alpha_{5,0.25}FOOD_{AC_{i,t}} + \alpha_{6,0.25}FOOD_{AV_{i,t}} + \alpha_{7,0.25}FOOD_{ST_{i,t}} + \alpha_{8,0.25}FOOD_{UT_{i,t}} + \alpha_{9,0.25}C02 + \mu_{0.25,i,t} \quad (2)$$

$$Q_{0.50}(GDPG_{it}) = \gamma_{0.50} + \alpha_{1,0.50}LAB_{i,t} + \alpha_{2,0.50}CAP_{it} + \alpha_{3,0.50}SSE_{i,t} + \alpha_{4,0.50}FOOD_{W_{i,t}} + \alpha_{5,0.50}FOOD_{AC_{i,t}} + \alpha_{6,0.50}FOOD_{AV_{i,t}} + \alpha_{7,0.50}FOOD_{ST_{i,t}} + \alpha_{8,0.50}FOOD_{UT_{i,t}} + \alpha_{9,0.50}C02 + \mu_{0.50,i,t} \quad (3)$$

$$Q_{0.75}(GDPG_{it}) = \gamma_{0.75} + \alpha_{1,0.75}LAB_{i,t} + \alpha_{2,0.75}CAP_{it} + \alpha_{3,0.75}SSE_{i,t} + \alpha_{4,0.75}FOOD_{W_{i,t}} + \alpha_{5,0.75}FOOD_{AC_{i,t}} + \alpha_{6,0.75}FOOD_{AV_{i,t}} + \alpha_{7,0.75}FOOD_{ST_{i,t}} + \alpha_{8,0.75}FOOD_{UT_{i,t}} + \alpha_{9,0.75}C02 + \mu_{0.75,i,t} \quad (4)$$

$$Q_{0.90}(GDPG_{it}) = \gamma_{0.90} + \alpha_{1,0.90}LAB_{i,t} + \alpha_{2,0.90}CAP_{it} + \alpha_{3,0.90}SSE_{i,t} + \alpha_{4,0.90}FOOD_{W_{i,t}} + \alpha_{5,0.90}FOOD_{AC_{i,t}} + \alpha_{6,0.90}FOOD_{AV_{i,t}} + \alpha_{7,0.90}FOOD_{ST_{i,t}} + \alpha_{8,0.90}FOOD_{UT_{i,t}} + \alpha_{9,0.90}C02 + \mu_{0.90i,t} \quad (5)$$

Where:
LAB = Labor force growth rate(Annual growth rate, Percentage).
CAP = Gross Fixed Capital Formation (Annual growth rate, Percentage).
Food W = Waste (Solid Food Waste CH4 emission Gigagrams.
Food AC = Food Access (Prevalence of Undernourishment).
Food AV = Food Availability (Average Protein Supply).
Food ST = Food Stability (Per Capita Food Production Variability).
Food UT = Food Utilization (People using at least basic drinking water services).
CO2 = Carbon Dioxide Emission (Environmental Quality).

5. Data and methodology

To achieve the objectives of this study, we have obtained panel data for low-income, lower-middle-income, and upper-middle-income countries from World

Development Indicators (WDI) and Food and Agriculture Organization (FAO) over the period 1990–2021. We have used the Methods of Moments Quantile Regression (MMQR) technique to probe the effect of food wastage, food security, and environment on the economic growth of developing countries.

6. Results and discussions

This section explains the results in detail. It includes summary statistics of key variables.

6.1 Descriptive statistics of key variables

Table 1 demonstrates the descriptive statistics of key variables for low-income, middle-income, and upper-middle-income countries. It can be observed from **Table 1** that in lower-income countries, all variables exhibit positive skewness except GDPG while in lower-middle income and upper-middle income countries, variable i.e., GDPG and FOOD-UT show negative skewness, and all other variables indicate positive skewness. FOOD-AV, SSE, and FOOD-AV have high tails due to their very low kurtosis.

6.2 Diagnostic tests

Our estimation strategy follows three steps. Firstly, we have tested the slope homogeneity/heterogeneity test. In this regard, two tests are common: (i) the Delta test presented by Pesaran and Yamagata, 2008; (ii) the HAC Robust Delta test or Delta Adjusted test by Blomquist and Westerlund, 2013.

Table 2 presents the results of the homogeneity/heterogeneity test for low-income, middle-income, and upper middle-income countries. Findings indicate that the slope is heterogeneous for all countries in both tests.

In the residuals, models of panel data can also indicate the cross-section dependence that may occur due to the spatial spillover effects, common shocks, or general error interdependence. It is assumed that there is cross-sectional independence both in causality and cointegration analysis. So, in the second step, we tested the dependence/independence of cross sections. For this purpose, we have applied the Pesaran CD Test.

Results of dependence/independence of cross sections through the Pesaran CD test in **Table 3** encapsulate that error terms of cross sections are corelated. It means that cross sections are not interdependent so there exists cross-section dependence in panel data of all countries.

In the third step, we have adopted another test known as the Westerlund test that has more explanatory power in cross-section dependence, and it assumes two hypotheses: (i) no cointegration means no long-run relation exists; (ii) cointegration exists (long-run relation). Westerlund test is divided into two dimensions, namely group dimension and panel dimension.

Table 4 shows the results of the Westerlund test, which elaborate that a long-run relationship exists as we have rejected the null hypothesis of no-cointegration.

Var	Mean	Median	Max	Min	Std. Dev.	Skewness	Kurtosis	JB	Prob.
Low-income countries									
GDPG	4.66	4.92	14.98	−6.12	4.35	−0.07	2.69	0.61	0.74
LAB	5,134,584	4,455,550	23,187,587	538,161	3,662,287	2.66	13.29	715.7	0.00
CAP	14.50	6.83	507.95	−36.12	49.33	7.98	79.65	32,689	0.00
SSE	17.10	12.52	51.03	5.29	11.16	1.27	3.89	38.67	0.00
FOOD_W	41.42	35.87	136.54	0.00	35.54	0.99	3.43	22.09	0.00
FOOD_AC	31.13	28.95	60.60	5.30	13.28	0.33	2.49	3.63	0.16
FOOD_AV	56.97	53.00	81.00	30.00	13.13	0.10	2.02	5.37	0.07
FOOD_ST	7.53	7.45	15.80	0.90	3.54	0.31	2.50	3.44	0.18
FOOD_UT	51.24	47.95	88.30	27.10	12.52	0.63	2.90	8.46	0.01
CO2	0.10	0.07	0.36	0.00	0.08	1.70	5.27	89.21	0.00
Lower middle-income countries									
GDPG	4.10	4.42	15.33	−21.30	3.92	−1.71	11.37	1078.8	0.00
LAB	34,852,297	8,264,429	451,000,000	4.35	82,433,049	3.92	17.95	3763.7	0.00
CAP	7.235191	7.12	9.91	−14.51	556,642	17.72	315.00	13,023	0.00
SSE	49.47	46.41	92.88	5.28	21.19	0.09	2.08	11.73	0.00
FOOD_W	401.97	76.46	3200.32	0.00	762.13	2.51	8.19	687.47	0.00
FOOD_AC	20.26	18.80	54.40	5.10	10.78	0.49	2.57	15.15	0.00
FOOD_AV	62.19	59.00	94.00	44.00	11.75	0.85	2.66	40.05	0.00
FOOD_ST	6.99	5.60	38.90	1.10	5.24	2.26	10.26	966.87	0.00
FOOD_UT	72.13	76.00	95.60	29.10	15.18	−0.67	2.63	25.23	0.00
CO2	1.09	0.70	45.33	0.07	2.76	13.23	208.79	56,862	0.00

Var	Mean	Median	Max	Min	Std. Dev.	Skewness	Kurtosis	JB	Prob.
Upper middle-income countries									
GDPG	3.92	4.04	14.35	−19.70	3.95	−1.43	11.14	673.10	0.00
LAB	6,924,844	2,708,509	83,821,485	60,325	11,442,809	3.64	20.67	3299.8	0.00
CAP	7.27	4.81	111.40	−38.68	18.18	1.19	8.13	289.56	0.00
SSE	66.18	66.07	109.99	21.98	18.26	−0.28	2.64	3.92	0.14
FOOD_W	160.63	45.33	1979.36	1.36	314.92	3.52	16.04	1985.3	0.00
FOOD_AC	15.88	14.40	39.90	5.00	9.46	0.74	2.57	21.43	0.00
FOOD_AV	67.36	67.00	88.00	47.00	9.13	0.03	2.55	1.86	0.40
FOOD_ST	14.93	11.30	60.20	2.10	10.75	1.38	4.93	102.79	0.00
FOOD_UT	84.73	86.60	99.60	53.00	10.19	−0.98	3.63	38.09	0.00
CO2	1.77	1.50	6.79	0.43	1.06	1.65	6.57	214.47	0.00

Table 1.
Summary statistics of key variables.

Tests	**Low income**	**Lower middle income**	**Upper middle income**
Homogeneity/heterogeneity test			
Status	Coefficients (p-value)		
Δ Delta Test (Pesaran and Yamagata, 2008)	25.598*** (0.000)	9.554*** (0.000)	13.014*** (0.000)
HAC Robust Delta Test (Blomquist and Westerlund, 2013) $\tilde{\Delta}^{Adjusted}$	87.413*** (0.000)	42.435*** (0.000)	19.539*** (0.000)

Table 2.
Diagnostic tests.

Variables	$\mathbf{Pesaran}CD_{Statistics}$ $(\mathbf{p-value})$		
	Low income	**Lower middle income**	**Upper middle income**
GDPG	2.235*** (0.000)	1.803*** (0.000)	14.820*** (0.000)
LAB	9.00.8*** (0.000)	9.114*** (0.000)	4.253*** (0.000)
CAP	7.153*** (0.000)	4.825*** (0.000)	32.222*** (0.000)
SSE	8.721*** (0.000)	11.259*** (0.000)	12.893*** (0.000)
FOOD_W	3.322*** (0.000)	26.005*** (0.000)	16.803*** (0.000)
FOOD_AC	3.146*** (0.000)	29.193*** (0.000)	11.406*** (0.000)
FOOD_AV	4.850*** (0.000)	12.268*** (0.000)	92.823*** (0.000)
FOOD_ST	6.836*** (0.000)	19.223*** (0.000)	13.234*** (0.000)
FOOD_UT	8.126*** (0.000)	16.589*** (0.000)	18.685*** (0.000)
CO2	9.493*** (0.000)	14.286*** (0.000)	19.814*** (0.000)

*Note: ***, **, and * show significance level at 1%, 5%, and 10%*

Table 3.
Cross-section dependence/independence.

6.3 MMQR regression results

Table 5 explains the estimated results for low-income, lower-middle, and upper-middle-income countries through four quantiles, i.e., 25th, 50th, 75th, and 90th. We elaborate on the findings with respect to the income levels of developing countries such as low-income, lower middle, and upper middle-income countries. Variables such as labor force, gross fixed capital formation, and secondary school enrollment

Statistic	Low income	Lower middle income	Upper middle income
G_t	−4.925*** (0.000)	−8.877*** (0.000)	−2.221*** (0.000)
G_a	−1.169** (0.032)	−4.986** (0.032)	−6.414** (0.032)
P_t	−9.689*** (0.000)	−16.340*** (0.000)	−9.695*** (0.000)
P_a	−3.426*** (0.001)	−2.298*** (0.001)	−3.199*** (0.001)

*Note: ***, **, and * show significance level at 1%, 5%, and 10%*

Table 4.
Cointegration check Westerlund test.

have a positive and significant relationship with the economic growth of all developing countries but have a different magnitude in all quartiles such as: in low-income countries, labor force has high magnitude at 50th quartile, gross fixed capital formation at 50th, and SSE at 75th quartiles. In lower middle-income countries, fluctuations are high for labor force and gross fixed capital formation and have approximately the same effect of secondary school enrollment at all quartiles, and the same is true for upper middle-income countries. The positive effect of labor force and gross fixed capital formation inserts that the increasing labor force growth rate and fixed capital formation can regulate the production possibilities, which in turn raises economic growth. The positive impact of secondary school enrollment on economic growth may be a result of the enhancement of educational levels and technological know-how.

Food wastage shows a negative and significant relationship with economic growth in all quartiles and all countries, but the magnitude is high in the 25th quartile in low-income countries and approximately the same in the lower-middle-income and upper-middle-income countries. Food wastage may evolve poor regulatory capacity, which in turn reduces economic growth. We have estimated food security in four aspects, i.e., food access, food availability, food stability, and food utilization. Food access is measured in terms of the prevalence of undernourishment, which indicates a negative and significant impact on economic growth in all quartiles but in low income-countries magnitude is high in the 50th quartile and in lower-middle-income and upper-middle-income countries degree is high at 75th quartiles. This negative relationship may be because the undernourishment of food access reduces economic growth. The next feature of food security is food availability, and we have estimated it in terms of average protein supply. Food availability has a positive and significant association with economic growth in all three categories of countries, but the results show that in low-income countries the extent of its effect is high in the 50th quartile but in the lower middle and upper middle countries their degree of effectiveness is approximately the same in all quartiles elaborating that high supply of proteins can enhance the economic growth of developing countries. The other two aspects of food security are food stability and food utilization, which exert a positive and significant effect on economic growth in all developing countries. The intensity of the effects of food stability and food utilization is high in the 25th quartile in low-income countries and approximately the same in all other quartiles in lower-middle-income and upper-middle-income countries. The positive impact of food utilization endorses that the variability of per capita food production leads to enhancing economic growth while

Variables	Location	Scale	$Q_{0.25}$ Coefficients [Std.Error]	$Q_{0.50}$ Coefficients [Std.Error]	$Q_{0.75}$ Coefficients [Std.Error]	$Q_{0.90}$ Coefficients [Std.Error]
Low-income countries						
LAB	2.945	−0.052	1.258*** [0.000]	8.159*** [0.008]	3.169*** [0.000]	0.198*** [0.003]
CAP	−1.158	0.0445	0.361*** [0.005]	0.416*** [0.006]	0.985*** [0.007]	0.159*** [0.002]
SSE	−0.058	0.0152	0.085*** [0.024]	0.947** [0.058]	0.982** [0.034]	0.018* [0.074]
FOOD_W	−0.147	0.0619	−0.982*** [0.005]	−0.882*** [0.021]	−0.158* [0.107]	−0.369** [0.055]
FOOD_AC	−8.179	0.028	−5.661*** [0.000]	−8.729*** [0.002]	−6.905** [0.040]	−7.831** [0.041]
FOOD_AV	−0.026	0.017	9.811*** [0.001]	10.294*** [0.002]	0.011** [0.030]	0.050* [0.070]
FOOD_ST	−0.076	0.037	0.674*** [0.025]	0.423** [0.048]	0.027*** [0.014]	0.049* [0.081]
FOOD_UT	−0.176	−0.009	0.131*** [0.000]	0.021*** [0.000]	0.052**8 [0.000]	0.019*** [0.000]
CO2	−0.202	0.085	−0.038*** [0.000]	−0.015*** [0.000]	−0.030*** [0.000]	−0.034*** [0.000]
C	0.153	−0.164	0.001*** [0.257]	0.003*** [0.398]	−0.561* [0.147]	0.474** [0.046]
Lower middle-income countries						
LAB	4.050	−11.712	4.509* [0.078]	3.371* [0.089]	4.362*** [0.008]	9.743*** [0.002]
CAP	−0.180	1.140	0.507*** [0.009]	0.185* [0.080]	0.596** [0.030]	0.378* [0.400]
SSE	0.016	0.061	0.036*** [0.025]	0.020*** [0.000]	0.031*** [0.008]	0.027** [0.033]
FOOD_W	−0.098	−0.169	−0.031** [0.036]	−0.012*** [0.000]	−0.030*** [0.000]	−0.021*** [0.025]
FOOD_AC	0.027	0.007	−0.089* [0.080]	−4.045*** [0.000]	−7.369* [0.069]	−0.001** [0.036]
FOOD_AV	−0.136	−0.060	0.050* [0.070]	0.030*** [0.000]	0.015** [0.040]	0.045*** [0.000]
FOOD_ST	−0.105	−0.099	0.055** [0.030]	0.024*** [0.000]	0.049* [0.801]	0.067** [0.030]
FOOD_UT	0.029	0.008	0.097* [0.060]	0.041*** [0.000]	0.281* [0.250]	0.055*** [0.025]
CO2	0.196	0.147	−0.085** [0.050]	−0.020*** [0.000]	−0.103*** [0.025]	−0.020** [0.039]
C	−2.419	14.293	0.102*** [0.020]	0.069*** [0.000]	0.331*** [0.000]	0.620* [0.070]

Variables	Location	Scale	$Q_{0.25}$ Coefficients [Std.Error]	$Q_{0.50}$ Coefficients [Std.Error]	$Q_{0.75}$ Coefficients [Std.Error]	$Q_{0.90}$ Coefficients [Std.Error]
Upper middle-income countries						
LAB	−7.178	9.153	1.078*** [0.001]	6.490*** [0.000]	1.377*** [0.000]	6.387*** [0.000]
CAP	0.638	0.122	0.659*** [0.000]	0.287*** [0.025]	0.230*** [0.000]	0.241*** [0.000]
SSE	0.013	0.019	0.031*** [0.002]	0.025*** [0.022]	0.106*** [0.000]	0.020*** [0.000]
FOOD_W	−0.055	−0.214	−0.022*** [0.000]	−0.015*** [0.000]	−0.007*** [0.000]	−0.014*** [0.300]
FOOD_AC	0.013	0.009	−1.987*** [0.006]	−1.832*** [0.000]	−3.712* [0.800]	−0.001*** [0.000]
FOOD_AV	−0.125	−0.288	0.029*** [0.000]	0.043** [0.040]	0.005* [0.060]	0.036*** [0.000]
FOOD_ST	−0.128	−0.224	0.031*** [0.009]	0.043*** [0.000]	0.037*** [0.000]	0.035* [0.500]
FOOD_UT	0.113	0.027	0.026*** [0.001]	0.069* [0.800]	0.021*** [0.020]	0.020*** [0.000]
CO2	0.198	0.267	−0.066*** [0.003]	−0.027*** [0.030]	−0.042*** [0.000]	−0.027*** [0.000]
C	−7.694	−4.908	0.137 [0.025]	0.137 [0.080]	−0.489 [0.030]	0.158 [0.000]

*Note: ***, **, and * show significance level at 1%, 5%, and 10%*

Table 5.
Method of moment quartile regression analysis (MMQR).

the positive influence of food utilization ratifies the development of economic growth due to the improvement in basic drinking water services. Carbon dioxide emission puts a negative influence on economic growth in all countries, and surprisingly, the fluctuation level is approximately the same in all quartiles in low-income, lower middle, and upper middle-income countries posting on the fact that environmental degradation reduces economic growth through poor health of labor, pollution, and deforestation.

7. Conclusions and policy recommendations

This study aimed to examine the effect of food wastage, food security, and the environment on the economic growth of developing countries over the period 1990–2021 by applying Methods of Moments Quantile Regression. Findings indicate that labor force, gross fixed capital formation, secondary school enrollment, food availability, food stability, and food utilization have a positive effect on the economic growth of developing countries, but the intensity of their effect differs from country to country. Further, the evidence indicates a negative association of food waste, food accessibility, and CO_2 emission with GDPG. Additionally, we have found that the strength of the effect of food waste on economic growth in low-income countries is high as compared with lower middle and upper middle-income countries. Nevertheless, while human and physical capital are efficient policy instruments for enhancing economic growth, they may not be sufficient for development. Other supportive policies such as reduction in food waste and food access are also needed. Most developing countries are blessed with natural resources, so there is also a need for the proper management of sustainable resources that may support economic growth. Moreover, to reinforce the behavioral changes, there would be given environmental and food protection awareness for all the citizens.

Acknowledgements

We would like to thank FAO for providing the data. We do not have any funding for this project.

Conflict of interest

We declare no conflict of interest.

A. Appendix

A.1 Food losses in low-income countries (2010–2021)

Year	Average	Maximum	Minimum	Countries	Food losses
2010	4.06	53	0.1	Burundi, Chad, Democratic Republic of the Congo, Benin, Ethiopia, Gambia, Guinea, Afghanistan, Liberia, Madagascar, Malawi, Mali, Mozambique, Nepal, Niger, Guinea-Bissau, Rwanda, Senegal, Sierra Leone, Somalia, Zimbabwe, Togo, Uganda, United Republic of Tanzania, Burkina Faso	Sorghum, Wheat, maize, rice, Tomatoes, fonio, millet
2011	3.25	50	0.02	Burundi, Chad, Democratic Republic of the Congo, Benin, Ethiopia, Gambia, Guinea, Afghanistan, Liberia, Madagascar, Malawi, Mali, Mozambique, Nepal, Niger, Guinea-Bissau, Rwanda, Senegal, Sierra Leone, Somalia, Zimbabwe, Togo, Uganda, United Republic of Tanzania, Burkina Faso	Cocoa beans, Onions and shallots, barley, bananas, pineapples, Tangerines, mandarins
2012	3.09	22	0.08	Burundi, Chad, Democratic Republic of the Congo, Benin, Ethiopia, Gambia, Guinea,	Wheat, maize, rice, Sorghum, bananas, potatoes, millet

Year	Average	Maximum	Minimum	Countries	Food losses
				Afghanistan, Liberia, Madagascar, Malawi, Mali, Mozambique, Nepal, Niger, Guinea-Bissau, Rwanda, Senegal, Sierra Leone, Somalia, Zimbabwe, Togo, Uganda, United Republic of Tanzania, Burkina Faso	
2013	3.72	63	0.08	Burundi, Chad, Democratic Republic of the Congo, Benin, Ethiopia, Gambia, Guinea, Afghanistan, Liberia, Madagascar, Malawi, Mali, Mozambique, Nepal, Niger, Guinea-Bissau, Rwanda, Senegal, Sierra Leone, Somalia, Zimbabwe, Togo, Uganda, United Republic of Tanzania, Burkina Faso	Plantains and cooking bananas, Groundnuts, excluding shelled, Tangerines, mandarins, Wheat, maize, rice, millet, Lentils, dry, Other beans, green, Bambara beans, dry, Sunflower seed, Sesame seed
2014	3.39	50	0.02	Burundi, Chad, Democratic Republic of the Congo, Benin, Ethiopia, Gambia, Guinea, Afghanistan, Liberia, Madagascar, Malawi, Mali, Mozambique, Nepal, Niger, Guinea-Bissau, Rwanda, Senegal, Sierra Leone, Somalia, Zimbabwe, Togo, Uganda, United Republic of Tanzania, Burkina Faso	Groundnuts, Sesame seed, Sweet potatoes, yams, Wheat, maize, rice, Other cereals, Sorghum, millet
2015	3.64	43	0.01	Burundi, Chad, Democratic Republic of the Congo, Benin, Ethiopia, Gambia, Guinea, Afghanistan, Liberia, Madagascar, Malawi, Mali, Mozambique, Nepal, Niger, Guinea-Bissau, Rwanda, Senegal, Sierra Leone, Somalia, Zimbabwe, Togo, Uganda, United Republic of Tanzania, Burkina Faso	Groundnuts, excluding shelled, Sesame seed, Cassava, fresh, Wheat, maize, rice, Sorghum
2016	3.21	29.96	0.01	Burundi, Chad, Democratic Republic of the Congo, Benin, Ethiopia, Gambia, Guinea, Afghanistan, Liberia, Madagascar, Malawi, Mali, Mozambique, Nepal, Niger, Guinea-Bissau, Rwanda, Senegal, Sierra Leone, Somalia, Zimbabwe, Togo, Uganda, United Republic of Tanzania, Burkina Faso	Groundnuts, Sesame seed, Sweet potatoes, yams, Wheat, maize, rice, Sorghum, millet, oats
2017	3.36	59.48	0.1	Burundi, Chad, Democratic Republic of the Congo, Benin, Ethiopia, Gambia, Guinea,	Sunflower seed, Other cereals, fonio, Wheat, maize, rice, Sorghum

Year	Average	Maximum	Minimum	Countries	Food losses
				Afghanistan, Liberia, Madagascar, Malawi, Mali, Mozambique, Nepal, Niger, Guinea-Bissau, Rwanda, Senegal, Sierra Leone, Somalia, Zimbabwe, Togo, Uganda, United Republic of Tanzania, Burkina Faso	
2018	3.13	49.7	0.1	Burundi, Chad, Democratic Republic of the Congo, Benin, Ethiopia, Gambia, Guinea, Afghanistan, Liberia, Madagascar, Malawi, Mali, Mozambique, Nepal, Niger, Guinea-Bissau, Rwanda, Senegal, Sierra Leone, Somalia, Zimbabwe, Togo, Uganda, United Republic of Tanzania, Burkina Faso	Cow peas, millet, Barley, fonio, Cow peas, oats.
2019	2.72	9.94	0.1	Burundi, Chad, Democratic Republic of the Congo, Benin, Ethiopia, Gambia, Guinea, Afghanistan, Liberia, Madagascar, Malawi, Mali, Mozambique, Nepal, Niger, Guinea-Bissau, Rwanda, Senegal, Sierra Leone, Somalia, Zimbabwe, Togo, Uganda, United Republic of Tanzania, Burkina Faso	Sorghum, Wheat, maize, rice, Barley, oats, fonio
2020	2.70	8.48	0.06	Burundi, Chad, Democratic Republic of the Congo, Benin, Ethiopia, Gambia, Guinea, Afghanistan, Liberia, Madagascar, Malawi, Mali, Mozambique, Nepal, Niger, Guinea-Bissau, Rwanda, Senegal, Sierra Leone, Somalia, Zimbabwe, Togo, Uganda, United Republic of Tanzania, Burkina Faso	Sorghum, Wheat, maize, rice, Barley, oats, fonio
2021	2.97	7.3	0.32	Nepal	Wheat, maize, rice

A.2 Food losses in lower middle-income countries (2010–2021)

Year	Average	Maximum	Minimum	Countries	Food losses
2010	7.78	62.5	0.07	Cameroon, Sri Lanka, Angola, Georgia, Ghana, Kenya, Kyrgyzstan, Lesotho, Mauritania, Nicaragua,	Onions and shallots, green, Maize, Millet, Eggplants, Cassava, fresh, Potatoes, Sorghum, Soya beans,

Year	Average	Maximum	Minimum	Countries	Food losses
				Philippines, Ukraine, Moldova, Viet Nam, Nigeria, Tunisia, Sudan, Eswatini, Zambia, Ukraine, Pakistan, CÃ'te d'Ivoire, Bolivia, India, Bangladesh	Groundnuts, Cassava, fresh, Sweet potatoes, Groundnuts, shelled, rice, wheat, Refined sugar, Sunflower seed, Raw milk of cattle, Pears, Meat of cattle, fresh or chilled, Okra, apples, yams, Eggplants, Plantains and cooking bananas, Strawberries, Other tropical and subtropical fruits, Chillies and peppers
2011	3.74	50	0.01	Cameroon, Angola, Georgia, Ghana, Kenya, Indonesia, Kyrgyzstan, Lesotho, Mauritania, Nicaragua, Philippines, Ukraine, Moldova, Viet Nam, Nigeria, Tunisia, Sudan, Eswatini, Zambia, Ukraine, Pakistan, CÃ'te d'Ivoire, Bolivia, India, Bangladesh	Lettuce and chicory, Maize, Millet, Cassava, fresh, Potatoes, Sorghum, Soya beans, Groundnuts, Cassava, fresh, Sweet potatoes, Groundnuts, shelled, rice, wheat, Refined sugar, Sunflower seed, Raw milk of cattle, Pears, Meat of cattle, fresh or chilled, Plantains and cooking bananas, Strawberries
2012	4.33	50	0.02	Cameroon, Angola, Georgia, Ghana, Kenya, Indonesia, Kyrgyzstan, Lesotho, Mauritania, Nicaragua, Philippines, Ukraine, Moldova, Viet Nam, Nigeria, Tunisia, Sudan, Eswatini, Zambia, Ukraine, Pakistan, CÃ'te d'Ivoire, Bolivia, India, Bangladesh	Mangoes, guavas and mangosteens, Tomatoes, Maize, Millet, Cassava, fresh, Potatoes, Sorghum, Soya beans, Groundnuts, Cassava, fresh, Sweet potatoes, Groundnuts, shelled, rice, wheat, Refined sugar, Sunflower seed, Raw milk of cattle, Pears, Meat of cattle, fresh or chilled, Plantains and cooking bananas
2013	5.31	60	0.01	Cameroon, El Salvador, Angola, Georgia, Ghana, Kenya, Indonesia, Kyrgyzstan, Lesotho, Mauritania, Nicaragua, Philippines, Ukraine, Moldova, Viet Nam, Nigeria, Tunisia, Sudan, Eswatini, Zambia, Ukraine, Pakistan, CÃ'te d'Ivoire, Bolivia, India, Bangladesh	Plantains and cooking bananas, Tomatoes, Rape or colza seed, Maize, Millet, Cassava, fresh, Potatoes, Sorghum, Soya beans, Groundnuts, Cassava, fresh, Sweet potatoes, Groundnuts, shelled, rice, wheat, Refined sugar, Sunflower seed, Raw milk of cattle, Pears, Meat of cattle, fresh or chilled, Plantains and cooking bananas, Sheep, Rape or colza seed
2014	2.58	35.7	0.01	Cameroon, Angola, Georgia, Ghana, Kenya, Indonesia, Kyrgyzstan, Lesotho, Mauritania, Nicaragua, Philippines, Ukraine, Moldova, Viet Nam, Nigeria, Tunisia, Sudan, Eswatini,	Flour of cassava, Tomatoes, Rape or colza seed, Maize, Millet, Cassava, fresh, Potatoes, Sorghum, Soya beans, Groundnuts, Cassava, fresh, Sweet potatoes, Groundnuts, shelled, rice,

Year	Average	Maximum	Minimum	Countries	Food losses
				Zambia, Ukraine, Pakistan, CÃ'te d'Ivoire, Bolivia, India, Bangladesh, Uzbekistan	wheat, Refined sugar, Sunflower seed, Raw milk of cattle, Pears, Meat of cattle, fresh or chilled, Plantains and cooking bananas, Sheep, Other tropical and subtropical fruit, Seed cotton, unginned, Plantains and cooking bananas, Broad beans and horse beans, dry, Buckwheat
2015	3.45	35	0.01	Cameroon, Angola, Georgia, Ghana, Indonesia, Kyrgyzstan, Lesotho, Mauritania, Nicaragua, Philippines, Ukraine, Moldova, Viet Nam, Nigeria, Tunisia, Sudan, Eswatini, Zambia, Ukraine, Pakistan, CÃ'te d'Ivoire, Bolivia, India, Bangladesh, Uzbekistan, Timor-Leste	Tomatoes, Rape or colza seed, Maize, Millet, Cassava, fresh, Potatoes, Sorghum, Soya beans, Groundnuts, Cassava, fresh, Sweet potatoes, Groundnuts, shelled, rice, wheat, Refined sugar, Sunflower seed, Raw milk of cattle, Pears, Meat of cattle, fresh or chilled, Plantains and cooking bananas, Sheep, Other tropical and subtropical fruit, Linseed, Tangerines, mandarins
2016	4.03	25	0.01	Cameroon, El Salvador, Angola, Georgia, Ghana, Indonesia, Kyrgyzstan, Lesotho, Mauritania, Honduras, Nicaragua, Philippines, Ukraine, Moldova, Viet Nam, Nigeria, Tunisia, Sudan, Eswatini, Zambia, Ukraine, Pakistan, CÃ'te d'Ivoire, Bolivia, India, Bangladesh, Uzbekistan	Mushrooms and truffles, Tomatoes, Rape or colza seed, Maize, Millet, Cassava, fresh, Potatoes, Sorghum, Soya beans, Groundnuts, Cassava, fresh, Sweet potatoes, Groundnuts, shelled, rice, wheat, Refined sugar, Sunflower seed, Raw milk of cattle, Pears, Meat of cattle, fresh or chilled, Plantains and cooking bananas, Sheep, Other tropical and subtropical fruit, Linseed, Tangerines, mandarins
2017	3.08	19.69	0.01	Cameroon, El Salvador, Angola, Georgia, Ghana, Indonesia, Kyrgyzstan, Lesotho, Mauritania, Honduras, Nicaragua, Philippines, Ukraine, Moldova, Viet Nam, Nigeria, Tunisia, Sudan, Eswatini, Zambia, Ukraine, Pakistan, CÃ'te d'Ivoire, Bolivia, India, Bangladesh, Uzbekistan	Seed cotton, unginned, Tomatoes, Rape or colza seed, Maize, Millet, Cassava, fresh, Potatoes, Sorghum, Soya beans, Groundnuts, Cassava, fresh, Sweet potatoes, Groundnuts, shelled, rice, wheat, Refined sugar, Sunflower seed, Raw milk of cattle, Pears, Meat of cattle, fresh or chilled, Plantains and cooking bananas, Sheep, Other tropical and subtropical fruit, Linseed, Tangerines, mandarins

Year	Average	Maximum	Minimum	Countries	Food losses
2018	3.10	22	0.01	Cameroon, Angola, Georgia, Ghana, Indonesia, Kyrgyzstan, Lesotho, Mauritania, Honduras, Nicaragua, Philippines, Ukraine, Moldova, Viet Nam, Nigeria, Tunisia, Sudan, Eswatini, Zambia, Ukraine, Pakistan, CÃ’te d'Ivoire, Bolivia, India, Bangladesh, Uzbekistan	Sugar beet, unginned, Tomatoes, Rape or colza seed, Maize, Millet, Cassava, fresh, Potatoes, Sorghum, Soya beans, Groundnuts, Cassava, fresh, Sweet potatoes, Groundnuts, shelled, rice, wheat, Refined sugar, Sunflower seed, Raw milk of cattle, Pears, Meat of cattle, fresh or chilled, Plantains and cooking bananas, Sheep, Other tropical and subtropical fruit, Linseed, Tangerines, mandarins, Taro
2019	2.72	6.42	0.01	Cameroon, Angola, Ghana, Indonesia, Lesotho, Mauritania, Honduras, Nicaragua, Philippines, Ukraine, Moldova, Viet Nam, Nigeria, Tunisia, Sudan, Eswatini, Zambia, Ukraine, CÃ’te d'Ivoire, Bolivia	Broad beans and horse beans, Tomatoes, Rape or colza seed, Maize, Millet, Cassava, fresh, Potatoes, Sorghum, Soya beans, Groundnuts, Cassava, fresh, Sweet potatoes, Groundnuts, shelled, rice, wheat, Refined sugar, Sunflower seed, Raw milk of cattle, Pears, Meat of cattle, fresh or chilled, Plantains and cooking bananas, Sheep, Other tropical and subtropical fruit
2020	2.71	6.42	0.01	Cameroon, Angola, Ghana, Lesotho, Mauritania, Ukraine, Moldova, Viet Nam, Nigeria, Tunisia, Sudan, Eswatini, Zambia, Ukraine, CÃ’te d'Ivoire	Maize, rice, millet, wheat, sorghum, barley

A.3 Food losses in upper middle-income countries (2010–2021)

Year	Average	Maximum	Minimum	Countries	Food losses
2010	4.62	26.31	0.09	Maize, rice, Plantains and others, grapes, Strawberries, Other berries and fruits of the genus Vaccinium, Soya beans, Coffee, green, Beans, dry, lupins, sorghum, Meat of cattle boneless, fresh or chilled, Wheat and meslin flour, Rice, Milled, rye, quinoa, asparagus, cabbages, spinach, Cantaloupes and other melons, tomatoes, Pumpkins, squash and gourds,	Cuba, Azerbaijan, Kazakhstan, Fiji, Iran, Armenia, Guatemala, Mauritius, Peru, Romania, Russian Federation, Namibia, Russian Federation, South Africa, Botswana, Venezuela (Bolivarian Republic of), Serbia Botswana, South Africa, Malaysia, Jordan

Year	Average	Maximum	Minimum	Countries	Food losses
				Carrots and turnips, Green garlic, bran of maize	
2011	3.88	27.44	0.02	Maize, rice, Plantains and others, grapes, Strawberries, Other berries and fruits of the genus Vaccinium, Soya beans, Coffee, green, Beans, dry, lupins, sorghum, Meat of cattle boneless, fresh or chilled, Wheat and meslin flour, Rice, Milled, rye, quinoa, asparagus, cabbages, spinach, Cantaloupes and other melons, tomatoes, Pumpkins, squash and gourds, Carrots and turnips, Green garlic, bran of maize, Watermelons	Cuba, Azerbaijan, Kazakhstan, Fiji, Iran, Armenia, Guatemala, Mauritius, Peru, Romania, Russian Federation, Namibia, Russian Federation, South Africa, Botswana, Venezuela (Bolivarian Republic of), Serbia Botswana, South Africa, Malaysia, Jordan, Mexico
2012	4.49	50	0.09	Maize, rice, Plantains and others, grapes, Strawberries, Other berries and fruits of the genus Vaccinium, Soya beans, Coffee, green, Beans, dry, lupins, sorghum, Meat of cattle boneless, fresh or chilled, Wheat and meslin flour, Rice, Milled, rye, quinoa, asparagus, cabbages, spinach, Cantaloupes and other melons, tomatoes, Pumpkins, squash and gourds, Carrots and turnips, Green garlic, bran of maize, Watermelons, Oil palm fruit	Ecuador, Fiji, Azerbaijan, Kazakhstan, Fiji, Iran, Armenia, Guatemala, Mauritius, Peru, Romania, Russian Federation, Namibia, Russian Federation, South Africa, Botswana, Venezuela (Bolivarian Republic of), South Africa, Jordan
2013	11.32	49.07	0.11	Maize, quinoa, grapes, Strawberries, Other berries and fruits of the genus Vaccinium, Soya beans, Coffee, green, Beans, dry, lupins, sorghum, Meat of cattle boneless, fresh or chilled, Wheat and meslin flour, Rice, Milled, rye, quinoa, asparagus, cabbages, spinach, Cantaloupes and other melons, tomatoes, Pumpkins, squash and gourds, Carrots and turnips, Green garlic, bran of maize, Watermelons, Oil palm fruit, Other vegetables, fresh	Ecuador, Gabon, Azerbaijan, Kazakhstan, Armenia, Guatemala, Mauritius, Peru, Romania, Russian Federation, Namibia, Russian Federation, South Africa, Paraguay, Botswana, Venezuela (Bolivarian Republic of), South Africa, Jordan, Saint Lucia, Thailand, Turkey
2014	4.31	22.44	0.01	Wheat, Maize, rice, Soya beans, Coffee, green, Beans, dry, lupins, sorghum, Meat of cattle boneless, fresh or	Azerbaijan, Kazakhstan, Armenia, Guatemala, Romania, Russian Federation, Namibia, Russian Federation,

Year	Average	Maximum	Minimum	Countries	Food losses
				chilled, Wheat and meslin flour, Rice, Milled, rye, quinoa, asparagus, cabbages, spinach, Cantaloupes and other melons, tomatoes, Pumpkins, squash and gourds, Carrots and turnips, Green garlic, bran of maize, Watermelons, Oil palm fruit, Other vegetables, fresh, Natural honey	South, Serbia, Africa, Paraguay, Botswana, Venezuela (Bolivarian Republic of), South Africa, Jordan, Saint Lucia, Thailand, Turkey, The former Yugoslav Republic of Macedonia
2015	5.96	46.87	0.02	Tomatoes, Maize, rice, Soya beans, Coffee, green, Beans, dry, lupins, sorghum, Meat of cattle boneless, fresh or chilled, Wheat and meslin flour, Rice, Milled, rye, quinoa, asparagus, cabbages, spinach, Cantaloupes and other melons, Pumpkins, squash and gourds, Carrots and turnips, Green garlic, bran of maize, Watermelons, Oil palm fruit, Raw milk of goats	Costa Rica, Fiji, Azerbaijan, Guatemala, Guyana, Kazakhstan, Armenia, Jamaica, Guatemala, Romania, Russian Federation, Namibia, Russian Federation, South, Serbia, Africa, Paraguay, Botswana, Venezuela (Bolivarian Republic of), South Africa, Jordan, Saint Lucia, Turkey, The former Yugoslav Republic of Macedonia
2016	5.50	30.03	0.07	Papayas, Maize, rice, Soya beans, Coffee, green, Beans, dry, lupins, sorghum, Meat of cattle boneless, fresh or chilled, Wheat and meslin flour, Rice, Milled, rye, quinoa, asparagus, cabbages, spinach, Cantaloupes and other melons, Green garlic, bran of maize, Watermelons, Oil palm fruit, Raw milk of goats, pulses	Colombia, Ecuador, Fiji, Azerbaijan, Iran, Guatemala, Guyana, Kazakhstan, Armenia, Jamaica, Russian Federation, Namibia, Russian Federation, South, Serbia, Africa, Paraguay, Botswana, Venezuela (Bolivarian Republic of), South Africa, Jordan, Saint Lucia, Turkey, The former Yugoslav Republic of Macedonia
2017	5.51	45.8	0.06	Maize, rice, Soya beans, Coffee, green, Beans, dry, lupins, sorghum, Meat of cattle boneless, fresh or chilled, Wheat and meslin flour, Rice, Milled, rye, quinoa, Green garlic, bran of maize, Watermelons, Oil palm fruit, Raw milk of goats, pulses, taro	Fiji, Azerbaijan, Iran, Guatemala, Guyana, Kazakhstan, Armenia, Jamaica, Russian Federation, Namibia, Russian Federation, South, Serbia, Africa, Paraguay, Botswana, Venezuela (Bolivarian Republic of), South Africa
2018	4.23	30.03	0.22	Maize, rice, Soya beans, Coffee, green, Beans, dry, lupins, fresh or chilled, Wheat and meslin flour, Rice, Milled, rye, quinoa, Green garlic, bran of maize, Watermelons, Oil palm fruit, Raw milk of goats, pulses, taro, millet	Fiji, Azerbaijan, Iran, Guatemala, Guyana, Kazakhstan, Armenia, Jamaica, Russian Federation, Namibia, Russian Federation, South, Serbia, Africa, Paraguay, Botswana, South Africa, Armenia

Year	Average	Maximum	Minimum	Countries	Food losses
2019	2.49	11.1	0.29	Wheat, maize, sorghum, millet, barley	Jordan, Namibia, South Africa, Botswana
2020	2.29	6.35	0.5	Mize, Wheat, maize, sorghum, millet, barley	Namibia, South Africa, Botswana
2021	17.2	31.6	5.5	Other fruits, n.e.c	South Africa

Author details

Muhammad Ramzan Sheikh*, Neelam Asghar Ali and Asma Aslam
School of Economics Bahauddin Zakariya University, Multan Pakistan

*Address all correspondence to: ramzansheikh@bzu.edu.pk

References

[1] FAO. The State of Food and Agriculture. Rome: Food and Agriculture Organization of the United Nations; 1996

[2] FAO. Data. 2021. Available from: h ttp://www.fao.org/faostat/en/#data/QC

[3] Gitonga ZM, De Groote H, Kassie M, Tefera T. Impact of metal silos on households' maize storage, storage losses and food security: An application of a propensity score matching. Food Policy. 2013;**43**:44-55

[4] Conrad Z, Blackstone NT. Identifying the links between consumer food waste, nutrition, and environmental sustainability: A narrative review. Nutrition Reviews. 2021;**79**(3):301-314

[5] Kosseva MR. Introduction: Causes and Challenges of Food Wastage. Elsevier; 2013. pp. xv-xxiv

[6] Rutten M. What economic theory tells us about the impacts of reducing food losses and/or waste: Implications for research, policy and practice. Agriculture & Food Security. 2013;**2**(1):13

[7] Irfanoglu ZB, Baldos UL, Hertel T, van der Mensbrugghe D. Impacts of Reducing Global Food Loss and Waste on Food Security, Trade. GHG Emissions and Land Use; 2014

[8] Sandaruwani JRC, Gnanapala WAC. Food wastage and its impacts on sustainable business operations: A study on Sri Lankan tourist hotels. Procedia Food Science. 2016;**6**:133-135

[9] Saputro WA, Purnomo S, Salamah U. Study of food waste of farmers' households in Klaten to support food security. Anjoro: International Journal of Agriculture and Business. 2021;**2**(2): 58-64

[10] Tefera T, Kanampiu F, De Groote H, Hellin J, Mugo S, Kimenju S, et al. The metal silo: An effective grain storage technology for reducing post-harvest insect and pathogen losses in maize while improving smallholder farmers' food security in developing countries. Crop Protection. 2011;**30**(3): 240-245

[11] Tefera T, Kanampiu F, De Groote H, Hellin J, Mugo S, Kimenju S, et al. The metal silo: An effective grain storage technology for reducing postharvest insect and pathogen losses in maize while improving smallholder farmers' food security in developing countries. Crop Protection. 2011;**30**(3):240-245

[12] Ma J, Wu S, Shekhar NV, Biswas S, Sahu AK. Determination of physicochemical parameters and levels of heavy metals in food waste water with environmental effects. Bioinorganic Chemistry and Applications. 2020;**2020**

[13] Huang CH, Liu SM, Hsu NY. Understanding global food surplus and food waste to tackle economic and environmental sustainability. Sustainability. 2020;**12**(7):2892

[14] Carletto G, Zezza A, Banerjee R. Towards better measurement of household food security: Harmonizing indicators and the role of household surveys. Global Food Security. 2013;**2**(1): 30-40

[15] Garrone P, Melacini M, Perego A. Opening the black box of food waste reduction. Food Policy. 2014;**46**:129-139

[16] Rockström J, Steffen W, Noone K, Persson Å, Chapin FS III, Lambin E, et al. Planetary boundaries: Exploring the safe operating space for humanity. Ecology and Society. 2009;**14**(2)

[17] Rutten M, Nowicki P, Bogaardt M-J, Aramyan L. Reducing Food Waste by Households and in Retail in the EU; a Prioritisation Using Economic, Land Use and Food Security Impacts. 2013. p. 156

[18] Timmermans A, Ambuko J, Belik W, Huang J. Food losses and waste in the context of sustainable food systems. In: CFS Committee on World Food Security HLPE. 2014

[19] EP. How to Avoid Food Wastage: Strategies for a More Efficient Food Chain in the EU. Brussels: European Parliament; 2011

[20] Bimmoy ED. Interrelationship of food wastage and sustainable development: A review. International Journal of Qualitative Research. 2022;**1**(3):228-232

[21] Mobaseri M, Mousavi SN, Haghighi MHM. Environmental Effects of Food Waste Reduction Using System Dynamics Approach. Energy Sources, Part A: Recovery, Utilization, and Environmental Effects. 2021. pp. 1-19

[22] Chrobog CK. Wasted: Understanding the economic and social impact of food waste [doctoral dissertation]. 2014

[23] Cm J. Impacts of food wastage on economic growth. World Food Policy. 2022;**8**(1):118-125

[24] Fami HS, Aramyan LH, Sijtsema SJ, Alambaigi A. The relationship between household food waste and food security in Tehran city: The role of urban women in household management. Industrial Marketing Management. 2021;**97**:71-83

[25] Sawaya WN. Impact of food losses and waste on food security. In: Water, Energy & Food Sustainability in the Middle East. Cham: Springer; 2017. pp. 361-388

[26] Smith P. Delivering food security without increasing pressure on land. Global Food Security. 2013;**2**(1):18-23

[27] James SJ, James CJFRI. The food cold-chain and climate change. Food Research International. 2010;**43**(7): 1944-1956

[28] Scherhaufer S, Moates G, Hartikainen H, Waldron K, Obersteiner G. Environmental impacts of food waste in Europe. Waste Management. 2018;**77**: 98-113

[29] Elofse S. Food Waste in SA—The Magnitude, Cost and Impacts. CSIR; 2014

[30] Cakar B, Aydin S, Varank G, Ozcan HK. Assessment of environmental impact of FOOD waste in Turkey. Journal of Cleaner Production. 2020; **244**:118846

[31] Thyberg KL, Tonjes DJ. The environmental impacts of alternative food waste treatment technologies in the US. Journal of Cleaner Production. 2017; **158**:101-108

[32] Shafiee-Jood M, Cai X. Reducing food loss and waste to enhance food security and environmental sustainability. Environmental Science & Technology. 2016;**50**(16):8432-8443

[33] Malahayati M, Masui T. Impact of reducing food wastage to the environment and economics: A preliminary finding of Indonesia case. Chemical Engineering Transactions. 2021;**89**:67-72

Chapter 3

In Situ and *Ex Situ* Agricultural Waste Management System

Mohd Muzamil, Sehreen Rasool and Ummyiah H. Masoodi

Abstract

The transformation of agricultural wastes, either *in situ* or *ex situ* manner can help to ensure nutrient recycling, energy generation, preparation of animal feed, medicines, packaging material, substrate for mushroom cultivation, biofuel production and product formulations. The *in situ* methods of waste management are prioritized as the problems of collection and transportation from the source can be avoided. The *in situ* methods are slow and require land and labour. The conversion of agricultural waste into fuel and useful value-added products is gaining traction and demands utilization of appropriate technology. In this context, the technological dependence on *ex situ* methods is higher than *in situ* methods. The selection of the particular method depends on the type of waste, process employed and final product required. The remedial measures can lead towards a sustainable future in terms of Safeguarding of human health, protection of soil, conservation of aquatic ecosystem and beneficial soil microbes and pave the way towards a cleaner, healthier and eco-friendly environment and ambience.

Keywords: agricultural waste, *in situ*, *ex situ*, mitigation, environment

1. Introduction

The rise in the world population to 7.9 billion in 2021 and the prediction that it will surpass 9 billion mark by 2050 and 11 billion by 2100 [1] has prompted the scientific community to involve advanced breeding methods, integrated managemental approaches and technological interventions to enhance the productivity and production of major cereal, vegetable, fruit and root crops and dairy animals [2]. The push resulted in an increase in the global production of maize (*Zea mays L.*), wheat (*Triticum aestivaum*) and rice (*Oryza sativa*) to 757 Mt, 1137 Mt and 757 Mt from 216 Mha, 197 Mha and 165 Mha, respectively [3, 4]. There has been sharp increase in the dairy animals, vegetables, fruit and root crops in the same period. However, the increase in the yield coincided with generation of large quantities of agricultural wastes. Taking an example of India, which benefitted immensely from the green revolution of the 1960s in enhancing the production and productivity of major crops and generates more than 3000 million tonnes of organic waste [5] with 686 Mt gross crop residues [6]. Globally, the agricultural sector is expected to contribute 4 billion tonnes of biomass/waste by 2050 [7]. The waste generated from the agricultural sector and agro-industries is rich in nutrients and cannot be left unprocessed or untreated

[8]. However, the management and disposal are cumbersome [9] owing to its large volume and immeasurable quantity [10]. Earlier, the focus was more towards production and productivity, thereby the agricultural wastes were neglected and discarded in sanitary landfills, decaying, burned or dumped in aquatic waterbodies.

The agricultural waste is classified into different categories depending on the areas from which it is generated, **Figure 1** [11]. Categorized agricultural (biomass) residues into primary, secondary and tertiary. The primary residues are produced in the process of plantation of food crops, secondary residues are released as by-products in the processing of food crops. The tertiary resides are generated when the biomass-based products are consumed by the human or animals. The traditional system of agricultural waste management is unsustainable owing to its hazardous consequences on human health [12], environment [13], soil microbes [14], water bodies [15] and global warming [13]. Several physical, chemical, biological and technological methods were employed to ensure the efficient agricultural waste management; however, most of them are infested with low degradation rates, labour intensive, costly and abysmal for environment [16]. All the events have converged and forced the policymakers to find the sustainable solution for the agricultural waste management.

2. Sustainable agriculture

The agricultural wastes can serve as the basic input for bio-economy with thrust towards health security, transformation of wastes into value-added products, livelihood security of farmers, job opportunities for youth and sustainability. As the world is shifting towards sustainable agriculture, agricultural waste management is being prioritized owing to the hazardous consequences on environment, health and economy [17] and prospective applications in the development of value-added products and services [8]. The proper waste management is one of the pillars of sustainable agriculture envisaged in sustainable development goals (SDG) of the United Nations. A number of methods were employed to transform the agricultural waste into useful products. These methods can be placed under two broad headings: *in situ* and *ex situ* waste management, **Figure 2**. The two methods can be clubbed together under the 3R (reduce, reuse and recycle) of integrated waste management system, **Figure 3**. Zaman and Lehmann [18] highlighted six wave innovation theory that took the world towards waste to wealth (energy) technologies, The sixth wave is related to zero waste in administration and manufacturing, emissions, product life and toxic use [20]. The agricultural waste management is not the prerogative of farmers (waste producers) only, but there are

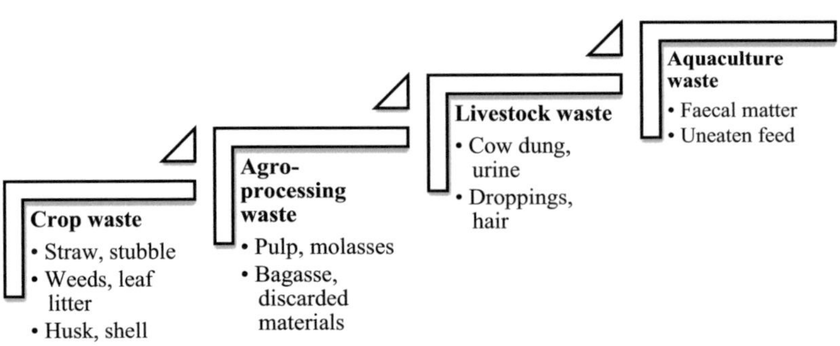

Figure 1.
Classification of agricultural waste [7].

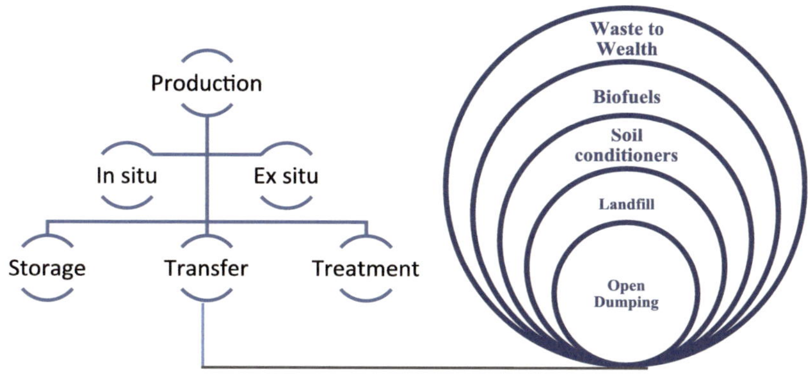

Figure 2.
Agricultural waste management functions [18, 19].

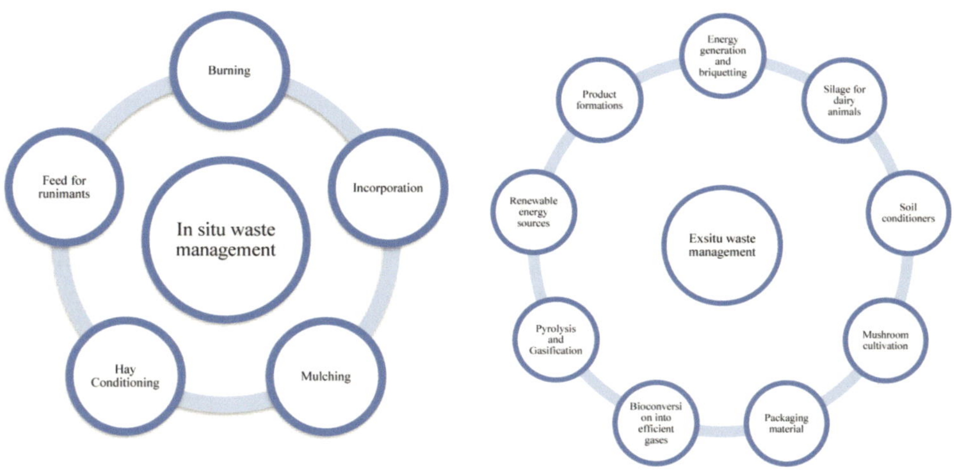

Figure 3.
Integrated waste management system.

multiple stakeholders that can play a meaningful role in its transformation. In fact, the policymakers are framing legislation and pushing hard to use the wastes for income augmentation, alternative energy option and basic input for bioeconomy.

2.1 *In situ* agricultural waste management

The term '*in situ*' refers to the management of the agricultural waste at the point of generation or production site. The method is prioritized as the cost involved in collection and transportation can be avoided. The *in situ* method is usually used for the nutrient recycling, bolstering of nutrient composition of the soil, soil and water conservation and animal feed. It includes

2.1.1 Burning

The use of the harvesters and combines often results in the generation of loose straws and stubbles. This waste along with the waste generated from unit agricultural operations from seedbed preparation to post-harvest processing is often burned in the open field to clear the land in shortest possible time. It is also believed to be an effective method of controlling the pest infestation from previous crop and weed control. There is a wide perception that the ash content improves the physical and chemical characteristics of the soil in the short term. The burning of the waste is preferred in areas where there is a technological vacuum or lack of alternatives or awareness, **Figure 4**. However, in the long run, there are dangerous consequences that have invited global concerns with respect to visibility [21], health [22], global climate change [23], release of polycyclic aromatic hydrocarbons [24] polychlorinated dibenzodioxins and polychlorinated dibenzofurans [11] and serious traffic accidents [25]. The pollutants released during biomass burning can produce significant changes in blood parameters as indicated by lymphocytosis, eosinophilia and neutrophilia in sheep [26] oxidative stress [27], as well as kidney [28] and liver dysfunctions in buffaloes [29]. Burning also releases considerable amounts of toxicants, which function as endocrine disruptors and affect the integrity of reproductive function in mammals and ultimately contribute to infertility [30].

In spite of the global condemnations, the agricultural waste burning is still prevalent in many developing countries. The emergence of Asian Brown Cloud (ABC) over South-East Asia is perceived as the direct consequence of burning the agricultural waste in the neighbouring countries. In order to avoid the burning of left-out stubbles and loose straw, the technocrats and scientific community have involved zero till drills and happy seeders to sow the seeds directly in the combined harvested fields. It has also helped to conserve more than 30% of the energy invested in seedbed preparation (tillage). This implies that there will be no compulsion to clear the field for the sowing of the next crop. It can help to curb the menace of burning agricultural fields on a large scale.

2.1.2 Incorporation

The agricultural wastes are lignocellulosic, with cellulose, hemicellulose and lignin as basic ingredients. In fact, cellulose accounts for 30–50% of the total biomass and

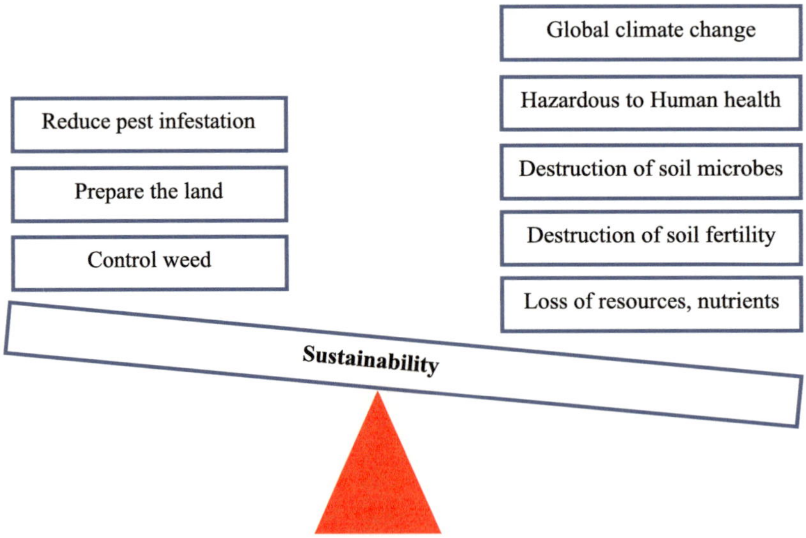

Figure 4.
Pros and cons of burning agricultural waste.

can be used for microbial transformation [31]. The agricultural waste generated on the field can be mixed with the soil and allowed to decay and degrade in presence of soil microbes. The microbial growth produces cellulose, endoglucanase and cellobiohydrolase enzymes to reduce the complex structure of cellulose into monomers. The decayed humus increases the water absorption capacity and bolsters the soil fertility status, directly or indirectly influencing the productivity of the crop [12]. However, the incorporation often hinders the movement of agricultural machinery as the straw gets accumulated at the furrow openers and seed tubes of seed-sowing machinery. Also, it is difficult to mix the loose straw and stubbles owing to their flexibility, mechanical strength and resilience [32]. The lack of viable cost-effective technology for the economic nutrient recycling has rendered the system unsustainable [33]. The waste must be subjected to particle-side reduction to increase its surface area. The small particles can be easily mixed with the soil and the higher surface area can provide feasible conditions for the soil microbes to degrade the waste in short duration. It is well documented that the particle size in the range of 1.27–7.62 cm can be helpful in the conversion of agricultural waste into nutrient-rich humus. In certain situations, the fungal or bacterial consortium is added to enhance the degradation rate of the agricultural waste. Muzamil et al. [12] developed a tractor-powered straw chopper with an ability to spray recommended dose of microbial culture to prepare the compost in shortest possible time. The paddy straw chopper was able to reduce 63% of the particles lower than 5 cm in size, **Figure 5**.

2.1.3 Mulching

The ability to conserve the soil fertility and judicious application of water is perceived as the backbone to achieve the sustainability of the agricultural system. This is more significant for the areas where the atmospheric conditions are harsh. The agricultural waste is spread between the rows and plants to conserve the soil, preserve the soil moisture and control the weed growth. As the plants are biodegradable, they get dissolved in the soil at the later phase and contribute to the soil fertility status. The current trend is to use plastic mulching with different colours to check the impact on pest infestation and productivity of crops. These plastic mulches are laid with the help of manual, engine-operated, tractor-drawn and self-propelled mulchers with an ability to spread drip irrigation channels below the mulched layer.

Figure 5.
Size reduction and spraying microbial culture—Image reproduced from Ref. [12].

2.1.4 Hay conditioning

The agricultural grass is harvested by means of mowers at a height of 3–10 cm from the ground. The mowers can be engine-operated, tractor operated or self-propelled. The harvested grass is spread uniformly over the surface by means of tractor-operated tedders. These tedders are used to turn the grass in order to ensure faster curing (drying). The dried grass is collected in the form of windrows using tractor-operated rakes. The windrow of grass is then compressed into high-density bales to decrease its volume and increase its weight. The bales are then transported and stored for future use. The process of mowing, tedding, raking and baling are the sub-constituents of '*hay conditioning*'.

2.1.5 Feed for ruminants

The harvested field often serves as the green pastures for ruminants. The animals are allowed to eat the agricultural wastes and the extract serves as the manure for the next crop. However, the trampling of the field by the ruminants is considered the potential source of sub-soiling. Sub-soiling is the process of hardening of soil surface at some depth, restricting the movement of water, runoff and washing off the upper soil fertile layer. The hard surface is known as plough sole or hard pan. The hard pan is often destroyed by means of sub-soiler (45–100 cm or even more) or chisel plough (25–35 cm). Sub-soiler possesses a single standard of large depth while chisel plough contains many standards of small depth. It is usually recommended to use a sub-soiler or chisel plough after 4–5 years of agricultural activities.

2.2 *Ex situ* agricultural waste management

Ex situ management differs from the *in situ* management in that the agricultural waste is collected and processed at separate facility. The cost of collection, transportation, storage and processing differentiates it from *in situ* agricultural waste management. *Ex situ* agricultural waste management can have many applications:

2.2.1 Energy generation and briquetting

The agricultural waste possesses huge volumes and is usually flexible [32]. The loose biomass with 100–200 kgm^{-3} density is compressed and transformed into densified blocks of 1200–1300 kgm^{-3} density briquettes. The biomass densification demands high pressure, provided with the help of piston-ram and screw-press machines. Moreover, the preparation of briquettes can be either binderless (no external additive is added and pressure is sufficient to bond) or with binder (external binders such as sodium bentonite, sand and molasses is added). In binderless process, high temperature (200°C) and pressure (1400 MPa) break the bonds and turn the cellulose into a fluid-like binding agent.

These briquettes are used in dairy processing units for animal feeding and in industries for the generation of energy. The energy can be in the form of steam (boiler), electricity or heat generation.

2.2.2 Silage for dairy animals

The process involves the conversion of agricultural wastes into silage (feed) for dairy animals and poultry. It includes cutting the harvested fodder (crop), compacting, storing and fermenting under an aerobic conditions in a silo. The agricultural wastes are usually low in nutrients and are mixed with agro-industrial by-products, such as rice bran, corn meal and coconut cake, to form complete feed (CF) for ruminants and poultry. The transformation into silage helps to maintain the feed nutrients in the forage for a long time. It is better than hay conditioning as it requires less space, fields can be prepared early and overcome the atmospheric constraints such as rain, snow and hailstorm. Moreover, the high-quality silage feed can be provided to the dairy animals and poultry around the year.

2.2.3 Soil conditioners

The wastes are collected from the field either manually or mechanically and converted into compost [34] and vermicompost [35].

2.2.3.1 Composting

The most common method of composting is bin-based and windrow-based composting. In bin-based method, a rectangular bin is used. The agricultural waste is mixed with 3–4 days-old cow dung and allowed to undergo degradation. The materials are mixed regularly to mix the water and microbes uniformly. This type of method is used for small-scale composting. At large scale, windrow composting is preferred in which the wastes are placed in long rows in the form of a windrow and 3–4 days old cow dung is mixed along with fungal consortium (*Aspergillus awamori and Trichoderma viridae*) to bolster the degradation process. The materials are mixed at regular intervals with the help of tractor operated compost turner-cum-mixer to distribute the microbes and moisture uniformly for rapid degradation of the agricultural waste [34].

2.2.3.2 Vermicomposting

The vermicomposting method is different from composting as it involves earthworms instead of bacteria or fungi in composting to degrade the agricultural waste. The involvement of earthworms is prioritized as it modifies the physical, chemical and biological properties of soil, resulting in quality manure production [36]. It is prepared either in rectangular bins [15] or smart vermicomposting bins [35] depending upon the application and quantity of waste. The fertility status of the vermicompost can help to cater to the needs of crop plants in order to promote the growth activity and serve as a sustainable alternative to chemical fertilizers [37]. It plays an essential role in the accretion of essential nutrients [38], enhancement of physio-chemical and biological properties [39, 40], improvement in porosity, aeration, drainage, solubility of the nutrients, water holding capacity [41, 42] and bioaccumulation of heavy metals in the soil [43]. In fact, the utilization of vermicompost-based organic agriculture can usher a revolution in terms of consumer health and environmental protection [44].

2.2.4 Mushroom cultivation

The mushroom cultivation requires substrates for the fungus to grow and transform into an edible mushroom. The lignocellulosic agricultural wastes are used as substrates and mediums for the microbes. These substrates are subjected to size reduction with the help of a chaff cutter or chopper. It is a case of nutrient recycling, where the nutrients in the agricultural wastes are extracted and utilized for the production of mushrooms. It is an ecologically sustainable and economically viable method of agricultural waste management [45, 46]. Moreover, the mushrooms prepared from banana stalks [47]; paddy straw [48], coffee husk [49] and agro-industrial waste [50] serve as essential ingredients to fight heart ailments and diabetes. The substrates are chopped, mixed with other ingredients and then placed in polythene bags of 5 kg or 10 kg before the addition of spawn (seeds). The spawn (seed) is added in two ways:

a. Spawn mixed with ingredients uniformly and filled in bags.

b. Spawn added in layers at the outer circumference of the polyethene bags.

In the first case, the spawn seeds are added at the time of mixing and then filled in polythene bags. There is a machine that fills one bag at one time. In the second case, the ingredients and spawn are added in layers. However, the spawn is placed at the outer circumference in layered structure. Usually, small holes are made in the polythene to allow the spawn seeds to germinate and emerge easily.

2.2.5 Packaging material

The agricultural waste, such as paddy straw, is also used as packing material for fruits and vegetables [51]. The farms and orchards are located at a far distance from the markets. Therefore, the fruits and vegetables are packed in cardboard or wooden boxes to transport them without any damage. The paddy waste is used for packing the perishable products due to its flexible and cushion characteristics. The packaging material can be made from tomato plants [52], starch from rice, wheat, potato peels and other agricultural wastes.

2.2.6 Bioconversion into efficient gases

Agricultural biomass can be transformed into solid (biochar and compost/vermicompost), liquid (bioethanol and biodiesel) or gaseous (producer gas, biogas and methane) or electricity through different technological interventions. The conversion of biomass into energy in anaerobic conditions with the active involvement of microorganisms is gaining traction [53]. The degradation of the agricultural biomass is accompanied by the liberation of methane and carbon dioxide, known as biogas. Biogas contains methane, carbon dioxide, hydrogen sulphide, ammonia, nitrogen and carbon monoxide with a calorific value of 4500 $kCalm^{-3}$. Biogas can be used for the generation of heat, electricity and engine operation. There are four processes involved in the preparation of biogas: hydrolysis, acetogenesis, acidogenesis and methanogenesis, **Table 1** Hydrolysis involves the breakdown of complex molecules into monomers. At the end of hydrolysis, acetic acid (CH_3COOH) is formed. The acetic acid is reduced to acids, either propionic acid or butyric acid. These acids are acted upon by methanogens and converted into biogas ($CH_4 + CO_2$). However, the type and composition of agricultural biomass [55], pH, density and degradation period [56] influence the quality and quantity of biogas, **Table 2**.

Biogas is prepared in biogas plants, which may be either fixed type or floating type. In fixed type biogas plant, the degradation chamber and drum are fixed. The output of the biogas plant is obtained from the top of the biogas plant. The degraded slurry is

Process	Chemical reaction	End product
Hydrolysis	Cellulose + H_2O Proteins + H_2O	Sugars Amino acids
Acidogenesis	$C_6H_{12}O_6 \rightarrow 3CH_3COOH$ $C_6H_{12}O_6 + 2H_2 \rightarrow 3CH_3CH_2COOH + 2H_2O$	Acetic acid Propionic acid
Acetogenesis	$CH_3CH_2COO^- + 3H_2O \rightarrow CH_3COO^- + HCO_3^- + H^+ + 3H_2$ $CH_3CH_2COO^- + 2H_2O \rightarrow 2CH_3COO^- + H^+ + 2H_2$ $4H_2 + 2HCO_3^- + H^+ \rightarrow CH_3COO^- + 4H_2O$	Acetate Hydrogen
Methanogenesis	$4CH_3COOH \rightarrow 4CO_2 + 4CH_4$ $CO_2 + 4H_2 \rightarrow CH_4 + 2H_2O$ $4CH_3OH + 6H_2 \rightarrow CH_4 + 2H_2O$	Methane Carbon dioxide Water

Table 1.
Chemical reactions involved in the preparation of biogas from biomass [54].

Substrate	Biogas potential nm^3/ mg fresh matter	Operations	Biogas requirement	Reference
Cow manure	63.3 ± 7.5	Lighting	0.127 m^3 per person	[57]
Poultry litter	241.6 ± 21.3	Cooking	0.227 m^3 per person	[58]
Agricultural waste (Straws, cobs and stalks)	124.4 ± 4.9	Power generation	0.425 m^3 per hp for engine operation	[58]
Vegetable, fruit and legume waste	158.1 ± 18.7	Slurry as organic fertilizer	-	[59]
Oil press residues	301.0 ± 9.3	-	-	[60, 61]

Table 2.
Biogas production and requirements from different substrates and activities.

collected from the bottom. In floating type biogas plant, the floating drum is placed over the top of degradation chamber. When the biogas starts pushing the floating drum, it moves upwards. The slurry is obtained in a similar manner as that of fixed type biogas plant. The biogas collected can be used for lighting, cooking and engine operation as fuel. The generation of the biogas from agricultural wastes can help to lower the greenhouse gases responsible for global warming. As per one estimate, the households utilizing biogas produce 50% lower greenhouse gas emissions than non-biogas households [62].

2.2.7 Pyrolysis and gasification

It is the process of burning the agricultural biomass (waste) at 350–700°C in controlled conditions to turn it into enriched products. The pyrolysis process converts the biomass into solid (char), liquid (bio-oil) and gaseous (fuel gas) mixture depending on the type of process, **Figure 6**. Biochar contains carbon and is an essential ingredient to increase the nutrient composition of the soil. When the temperature exceeds more than 700°C, it reduces the biomass into hydrogen and carbon monoxide-based gaseous fuel. This gas is known as producer gas with individual constituents as 18–22% carbon monoxide, 15–20% hydrogen, 1–5% methane, 8–12%

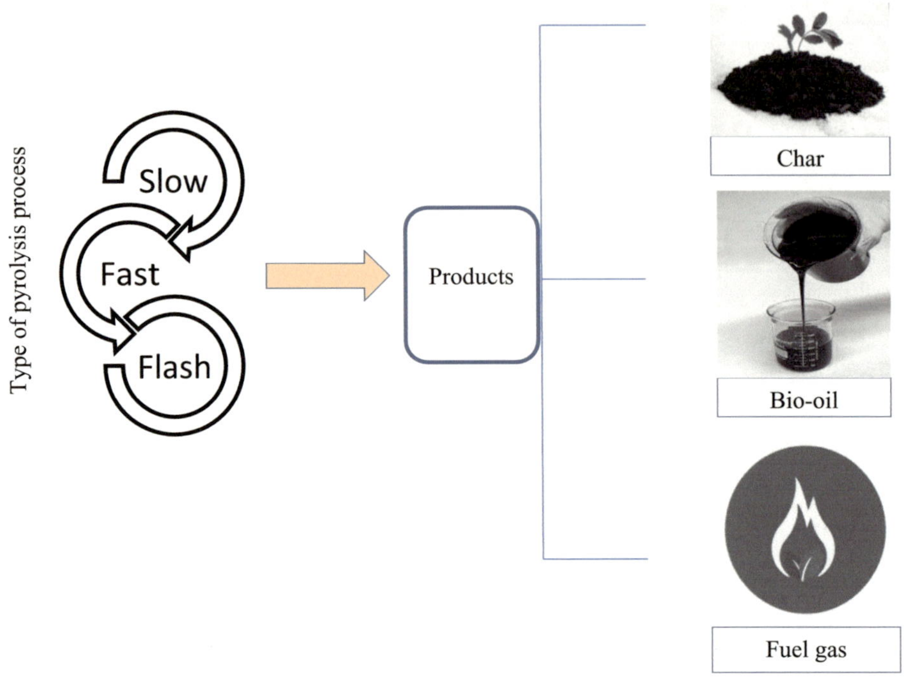

Figure 6.
Products formed by different methods of pyrolysis [63].

carbon dioxide and 45–55% nitrogen, **Table 3**. The calorific value of producer gas is 4.2–5.0 MJ/Nm^3 with conversion efficiency of 80%. There are four types of gasifiers, depending on the type of substrate and application. The efficiency of the burning of producer gas is 80%, higher than biogas (60%) and liquified petroleum gas (60%). The quality of the gas produced depends on the type of substrate, temperature, bed height and gasifying agent. Moreover, hydrogen is also emerging as a potential energy source. Ahmad et al. [65] reported that gasification is the most efficient technique for the production of hydrogen gas from biomass.

2.2.8 Renewable energy sources

The world is under tremendous pressure to seek alternate sources of fuel after the emergence of greenhouse gases led to global warming, increase in energy consumption

Type of reaction	Chemical interaction	Reference
Oxidation reaction	$C + O_2 \rightarrow CO_2$ $C + 1/2\ O_2 \rightarrow CO$	[64]
Boudouard	$C + CO_2 \rightarrow 2CO$	
Water gas	$C + H_2O \rightarrow CO + H_2$	
Methanation	$C + 2H_2 \rightarrow CH_4$	
Water gas shift	$CO + 2H_2O \rightarrow CO_2 + H_2$	

Table 3.
Chemical reactions involved in gasification.

[66] and growing concerns for health and environment from fossil fuels. There are four generations of biofuel production depending on the type of biomass, methods and technological procedure adoption, **Figure** 7. Earlier, the biofuels were manufactured from started based food materials such as corn, sugarcane, wheat, sorghum and millet. However, the controversy over food crops vs. non-food crops for biofuel production has rendered the system untenable. In the second generation, lignocellulosic biomass was utilized with the prediction that 442 billion litres of bioethanol (**Table 4**) can be produced from rice, corn, sugarcane and wheat straw [73]. Third generation exploited the potential of algae to produce biofuels. Fourth generation relies heavily on genetically engineered feedstock for the production of biofuels. The agricultural wastes [67] can be effectively used in the production of biofuels.

Biodiesel is prepared from biomass/vegetable oils through the process of transesterification and utilized as the substitute for conventional diesel in automobiles. Biodiesel is also known as free fatty acid alkyl ester. The vegetable oils are mainly composed of fatty acids of glycerol known as triglycerides. When the triglyceride molecule reacts with methanol, it produces fatty acid methyl ester (biodiesel) and glycerol. In general, transesterification is the reaction of fat or oil with the alcohol to produce esters and glycerine. Currently, biodiesel is mixed with conventional diesel in different proportions to assess its efficacy for various operations. The proportion of B5 signifies that 5% is biodiesel and 95% is conventional diesel fuel. Similarly, there are other proportions such as B10, B15 and B85. The involvement of biodiesel lowers the emissions of almost all the pollutants from the automobiles with the exception of nitrogen oxide (NO_2).

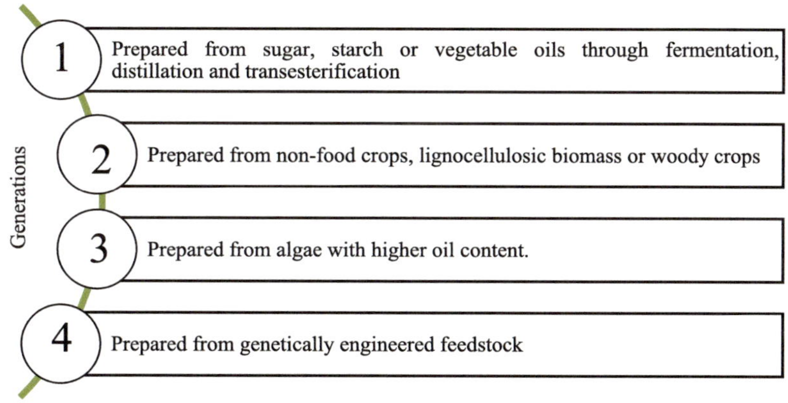

Figure 7.
Four generations of biofuel production.

Biofuel	Substrate	Reference
Bioethanol	Agricultural wastes and crop wastes	[67–69]
Bioethanol	Vegetable waste (potato, carrot and onion peel)	[70]
Bioethanol	Banana stem	[71]
Bioethanol	Agro-industrial waste	[72]

Table 4.
Agricultural wastes used for bioethanol production.

Finished product	Method	Reference
Aquatic and terrestrial Dal weed into vermicompost	Mechanical interface and technological intervention	[15, 32, 35, 36]
Paddy straw stubble into compost	Mechanical interface and microbial inoculum	[12, 20, 34, 81]
Utilization of biomass as fuel	Procedural methodology	[25]
Waste to value-added products	—	[26]
Waste for mushroom production	Technological intervention and standard method	[45, 47]
Waste for production of biogas, producer gas, hydrogen and electricity	Hydrolysis and digestion	[53, 55–57, 61, 64, 65]
Waste for biofuel production	Acid hydrolysis and fermentation	[67, 68, 70, 73]
Preparation of medicines and antioxidants from wastes	Standard procedures	[74, 75, 77]

Table 5.
Current methods used for agricultural waste management.

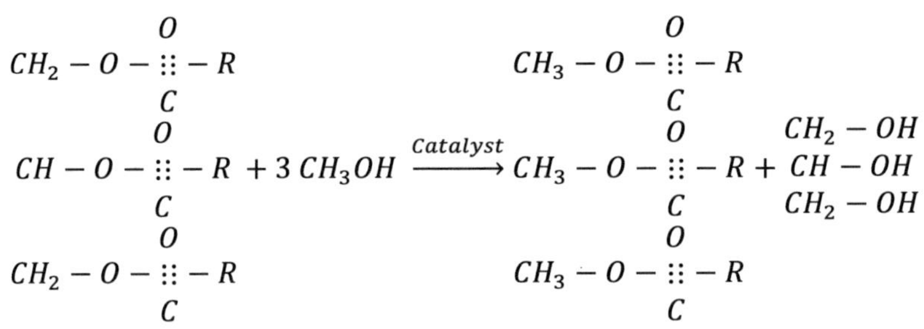

Triglyceride Methanol Fatty acid methyl ester Glycerol

2.2.9 Product formations

A number of products can be formed from agricultural wastes such as dishes and plates from bagasse (sugarcane) by-product; antioxidants (pineapple waste [74], orange peel) [75]; pharmaceutical products (fruit and vegetable peel) [76]; antibiotic oxytetracycline (corn cobs, sawdust and rice hulls) [77] and enzyme production [78]. The agricultural wastes can also be used to prepare indigenous fermented products such as Indonesian Oncom [79] and Indonesian and Malaysian tempeh [80].

Currently, a number of studies are diverted to involve the technical procedures, technological interventions and methodological processes of agricultural waste management, **Table 5**. All the methods are tested for technical feasibility, economic viability and commercial scalability to make it affordable to the main stakeholder—the farmer.

3. Conclusion

The agricultural wastes are burgeoning at a rapid rate and demand appropriate *in situ* and *ex situ* managemental strategy, depending on the type of substrate, area of generation, quantity and final product required. On certain occasions, the waste can be chopped and incorporated into the soil for nutrient recycling and easy movement of agricultural machinery [81]. The incorporated layer protects the soil from erosion, conserves the moisture and provides humus for the growth of the plants. The *in situ* methods are usually slow and time-consuming. However, at times, it becomes difficult to manage the agricultural waste at the source of its generation. In such cases, it is better to collect, transport and transform it into briquettes for energy generation, feeding material for ruminants and poultry, biofuel production, conversion into biogas and producer gas for household and industrial units. It can be also used for composting, vermicomposting-based entrepreneurship enterprises or packaging material for perishable agricultural products. The conversion of agricultural wastes is imperative to protect human health, environment and arrest global climate change. The 'waste to wealth' can ensure income augmentation of the farmers and sustainability of the agricultural system.

Acknowledgements

The authors would like to appreciate the help rendered by the authorities at SKUAST-K that finally led to the completion of the herculean task.

Conflict of interest

All the authors contributed equally to the preparation of the manuscript and have unanimously decided to publish it. Therefore, there is no conflict of interest whatsoever.

Author details

Mohd Muzamil[1*], Sehreen Rasool[2] and Ummyiah H. Masoodi[3]

1 College of Agricultural Engineering and Technology, SKUAST-K, Shalimar, Jammu & Kashmir, India

2 College of Temperate Sericulture, SKUAST-K, Mirgund, Jammu & Kashmir, India

3 Division of Vegetable Science, SKUAST-K, Shalimar, Shalimar, Jammu & Kashmir, India

*Address all correspondence to: muzamil4951@gmail.com

References

[1] Koops SHA, Leeuwen CJ. The challenges of water, waste and climate change in cities. Environment, Development and Sustainability. 2017; **19**:385-418

[2] Tripathi RP, Peeyush S, Surendra S. Soil physical response to multi-year rice-wheat production in India. International Journal of Soil Science. 2006;**1**:91-107

[3] FAOStat. FAO Stat. Rome: FAO; 2021

[4] Erenstein O, Jaleta, M, Sonder K, Mottaleb K, Prassana BM. Global maize production, consumption and trade: trends and R&D implications. Food Security. 2022;**14**:1295-1319. DOI: 10.1007/s12571-022-01288-7

[5] Sharholy M, Ahmad K, Mahmood G, Trivedi RC. Municipal solid waste management in Indian cities- A review. Waste Management. 2008;**28**:459-467

[6] Hiloidhari M, Das D, Baruah DC. Bioenergy potential from crop residue biomass in India. Renewable and Sustainable Energy Reviews. 2014;**32**: 504-512

[7] Duque-Acevedo M, Belmonte-Urena L, Cortes-Garcia FJ, Camacho-Ferre F. Agricultural waste: Review of the evolution, approaches and perspectives on alternative uses. Glob. Ecol. Conserv. 2020;**22**:e00902. DOI: 10.1016/j.gecco.2020.e00902

[8] Sadh PK, Duhan S, Duhan JS. Agro-industrial wastes and their utilization using solid state fermentation: A review. Bioresources and Bioprocess. 2018;**5**(1)

[9] Rodriguez-Couto S. Exploitation of biological wastes for the production of value-added products under solid-state fermentation conditions. Biotechnology Journal. 2008;**3**:859-870

[10] Wang B, Dong F, Chen M, Zhu J, Tan J, Fu X, et al. Advances in recycling and utilization of agricultural wastes in China: Based on environmental risk, crucial pathways, influencing factors, policy mechanism. The tenth international conference on waste management and technology (ICWMT). Procedia Environmental Sciences. 2016; **31**:12-17

[11] Li A, Antizar-Ladislao B, Khraisheh M. Bioconversion of municipal solid waste to glucose for bio-ethanol production. Bioprocess and Biosystems Engineering. 2007;**30**: 189-196

[12] Muzamil M, Mani I, Kumar A, Shukla L. An engineering intervention to prevent paddy straw burning through in situ microbial degradation. Journal of Institution of Engineers. 2021;**102**:11-17

[13] Kaab A, Sharifi M, Mobli H, Pelesaraei A, N. and Chau, K. Combines life cycle assessment and artificial intelligence for prediction of output energy and environmental impacts of sugarcane production. Science of the Total Environment. 2019;**664**:1005-1019

[14] Gupta PK, Sahai S, Singh N, Dixit CK, Singh DP, Sharma C. Residue burning in rice-wheat cropping system: Causes and implications. Current Science. 2004;**87**(**12**):1713-1715

[15] Kumar S. Aquatic weeds problems and management in India. Indian Journal of Weed Science. 2011;**43**(3-4):118-138

[16] Najar IAK. Vermicomposting of invasive species Azolla pinnata with

Eisenia foetida. An International Quartely Journal of Life Science. 2010; **5**(2):239-241

[17] Dhar H, Kumar S, Kumar R. A review on organic waste to energy systems in India. Bioresource Technology. 2017;**245**:1229-1237

[18] Zaman AU, Lehmann S. Urban growth and waste management optimization towards 'zero waste city'. City, Culture and Society. 2011;**2**:177-187

[19] USDA-NRCS. Agricultural waste management systems. Agricultural Waste Management Field Handbook. 2011;**651**(9):31

[20] Pietzsch N, Ribeiro JLD, Medeiros JF. Benefits, challenges and critical factors of success for zero waste: A systematic literature review. Waste Management. 2017;**67**:324-353

[21] Penner JE, Dickinson RE, Oneill CA. Effects of aerosol from biomass burning on the global radiation budget. Science. 1992;**256**:1432-1434

[22] Gadde B, Bonnet S, Menke C, Garivait S. Air pollutant emissions from rice straw open field burning in India, Thailand and the Philippines. Environmental Pollution. 2009;**157**: 1554-1558

[23] Lemieux PM, Lutesb CC, Santoiannib DA. Emissions of organic air toxics from open burning. Progress in Energy and Combustion Science. 2004; **30**:1-32

[24] Korenaga T, Liu X, Huang Z. The influence of moisture content on polycyclic aromatic hydrocarbons emission during rice straw burning. Chemosphere - Global Change Science. 2001;**3**:117-122

[25] Chou CS, Lin SH, Lu WC. Preparation and characterization of solid biomass fuel made from rice straw and rice bran. Fuel Processing Technology. 2009;**90**:980-987

[26] Ahmed WM, Abou Mossallam AA, Hanfiand EM, Shalaby SI. Observations on fertility and mutagenicity of Egyptian ewes exposed to products of rice straw combustion with special reference to the effect of flunixine meglumine. *Journal of the* Egyptian Veterinary Medical Association. 2003;**63**:195-212

[27] Rivero DH, Soares SR, Filho GL, Godleski JJ, Antonangelo L, Dolhnikoff M, et al. Acute cardiopulmonary alterations induced by fine particulate matter of São Paulo, Brazil. Toxicological Sciences. 2005;**85** (**2**):898-905

[28] Uboh FE, Akpanabiatu MI, Eteng M, Eyong EU. Exposure to gasoline and kerosene vapours: A risk factor for nephrotoxicity in rats. Internet Journal of Toxicology. 2010;**7**:2

[29] Massaguer A, Perez-Del-pulgar S, Engel P, Serratosa J, Bosch J, Pizcueta P. Concanavalin-A-induced liver injury is severely impaired in mice deficient in P-selectin. Journal of Leukocyte Biology. 2002;**72**:262-270

[30] Smith TL, Merry ST, Harris DL, Ford JJ, Ike J, Archibong AE, et al. Species specific testicular and hepatic microsomal metabolism of benzopyrene, an ubiquitous toxicant and endocrine disruptor. Toxicology In Vitro. 2007;**21**: 753-758

[31] Koul B, Yakoob M, Shah MP. Agricultural waste management strategies for environmental sustainability. Environmental Research. 2022;**206**:112285

[32] Muzamil M, Mani I, Kumar A, Lande S. Influence of moisture content, loading rate

and internode position on the mechanical properties of paddy and wheat straw. International Journal of Bio-resource and Stress Management. 2016;7(2):280-285

[33] Jeyabal A, Kuppuswamy G. Recycling of organic wastes for the production of vermicompost and its response in rice–legume cropping system and soil fertility. European Journal of Agronomy. 2001;**15**(3):153-170

[34] Muzamil M, Mani I, Shukla L, Lande S. Efficiency of pusa compost inoculant along with windrow turner for rapid degradation of paddy straw. Journal of Pure and Applied Microbiology. 2014;**8**(**1**):569-576

[35] Kumar A, Muzamil M, Dixit J. A smart vermicomposting bin for rapid degradation of Dal lake aquatic weed into fortified vermicompost. International Journal of recycling of organic waste in Agriculture. 2022;**12**(2). DOI: 10.30486/ijrowa.2022.1955835.1445

[36] Molina MJ, Soriano MD, Ingelmo F, Linares J. Stabilisation of sewage sludge and vinasse bio-wastes by vermicomposting with rabbit manure using Eisenia fetida. Bioresource Technology. 2013;**137**:88-97

[37] Chauhan HK, Singh K. Potancy of Vermiwash with Neem plant parts on the Infestation of Earias vittella (Fabricius) and Productivity of Okra (Abelmoschus esculentus) (L.) Moench. Asian Journal of Research and Pharmaceutical Science. 2015;**5**(1):36-40

[38] Gandhi A, Sundari SU. Effect of vermicompost prepared from aquatic weeds on growth and yield of eggplant (Solanum smelongena L.). Journal of Biofertilizer and Biopesticide. 2012;**3**(5): 1-4

[39] Ansari AA, Jaikishun S. Vermicomposting of sugarcane bagasse and rice straw and its impact on the cultivation of Phaseolus vulgaris L. in Guyana, South America. Journal of Agricultural Technology. 2011;7(2): 225-234

[40] Chauhan HK, Singh K. Effect of tertiary combinations of animal dung with agrowastes on the growth and development of earthworm Eisenia fetida during organic waste management. International Journal of Recycling Organic Agriculture. 2013;**2**:11

[41] Ismail SA. The Earthworm Book. Mapusa: Other India Press; 2005. p. 101

[42] Edwards CA, Subler S, Arancon N. Quality criteria for vermicomposts. In: Edwards CA, Arancon NQ, Sherman RL, editors. Vermiculture Technology: Earthworms, Organic Waste and Environmental Management. Boca Raton: CRC Press; 2011. pp. 287-301

[43] Hemalatha B. Vermicomposting of fruit waste and industrial sludge. International Journal of Advanced Engineering Technology. 2012;**3**(2): 60-63

[44] Kaplan M. The National Master Plan for Agricultural Development in Suriname. Final Report. Kaplan Planners Ltd. Regional and Environmental Planning; 2016. pp. 255. https://www.share4dev.info/kb/documents/5426.pdf

[45] Randive SD. Cultivation and study of growth of oyster mushroom on different agricultural waste substrate and its nutrient analysis. Advances in Applied Science Research. 2012;**3**: 1938-1949

[46] Chiu SW, Moore D. Threats to biodiversity caused by the traditional mushroom cultivation in China. In: Moore D. Nauta MM, Evans SE, Rotheroe M. editors. Fungal

conservation: Issues and Solutions, British Mycological Society Symposia, Chapter. Cambridge: Cambridge University Press; 2001. pp. 111-119

[47] Siqueira F, Martos E, Silva R, Dias E. Cultivation of Pleurotus sajor-caju on banana stalk and bahia grass-based substrates. Horticultura Brasileira. 2011; **29**:199-204

[48] Akinyele JB, Fakoya S, Adetuyi CF. Anti-growth factors associated with Pleurotus ostreatus in a submerged liquid fermentation. Malaysian Journal of Microbiology. 2012;**8**:135-140

[49] Murthy PS, Manonmani HK. Bioconversion of coffee industry wastes with white rot fungus Pleurotus florida. Research Journal of Environmental Sciences. 2008;**2**:145-150

[50] Babu PD, Subhasree RS. Valuing the suitable agro-industrial wastes for cultivation of *P. platypus* and *P. eous*. Advances in Biology Research. 2010;**4**: 207-210

[51] Valdés A, Mellinas AC, Ramos M, Garrigos MC, Jiménez A. Natural additives and agricultural wastes in biopolymer formulations for food packaging. Frontiers in Chemistry. 2014;**2**:6

[52] Van Crevel R. Bio-based food packaging in sustainable development. Challenges and opportunities to utilize biomass residues from agriculture and forestry as a feedstock for bio-based food packaging. Food and Agriculture Organization of the United Nations; 2001. pp. 1-70

[53] Liu Y, Zhu Y, Jia H, Yong X, Zhang L, Zhou J, et al. Effects of different biofilm carriers on biogas production during anaerobic digestion of corn straw. Bioresource Technology. 2017;**244**:445-451

[54] Nwokolo N, Mukumba P, Obileke K, Enebe M. Waste to energy: A focus on the impact of substrate type in biogas production. PRO. 2020;**8**:1224

[55] Mao C, Feng Y, Wang X, Ren G. Review on research achievements of biogas from anaerobic digestion. Renewable and Sustainable Energy Reviews. 2015;**45**:540-555

[56] Li K, Liu R, Cui S, Yu Q, Ma R. Anaerobic co-digestion of animal manures with corn stover or apple pulp for enhanced biogas production. Renewable Energy. 2018;**118**:335-342

[57] Ghmire PC. SNV supported domestic biogas programmes in Asia and Africa. Renewable Energy. 2013;**49**:90-94

[58] Roubík H, Mazancová J, Banout J, Verner V. Addressing problems at small-scale biogas plants: A case study from central Vietnam. Journal of Cleaner Production. 2016;**112**:2784-2792

[59] Rodolfo S, Anh LH, Konrad K. Feasibility assessment of anaerobic digestion technologies for household wastes in Vietnam. Journal of Vietnam Environment. 2016;**7**:1-8

[60] Bozzetto S, Pezzaglia M, Rossi L, Pecorino B. Considerazioni sul Potenziale del “Biogas Fatto Bene” Italiano Ottenuto Dalla Digestione Anaerobica di Matrici Agricole. Metodologia di Stima e Analisi dei Dati del Position Paper del Consorzio Italiano Biogas; Italian. Lodi, Italy: Consorzio Italiano Biogas (CIB); 2016

[61] Valli L, Rossi L, Fabbri C, Sibilla F, Gattoni P, Dale BE, et al. Greenhouse gas emissions of electricity and biomethane produced using the BiogasdonerightTM system: Four case studies from Italy. Biofuels, Bioproducts and Biorefining. 2017;**11**:847-860

[62] Zhang X, Luo L, Skitmore M. Household carbon emission research: An analytical review of measurement, influencing factors and mitigation prospects. Journal of Cleaner Production. 2015;**103**:873-883

[63] Lee SY, Sankaran R, Chew KW. Waste to bioenergy: A review on the recent conversion technologies. BMC Energy. 2019;**1**:4

[64] Ghani WAW, Moghadam RA, Salleh MAM, Alias AB. Air gasification of agricultural waste in a fluidized bed gasifier: Hydrogen production performance. Energies. 2009;**2**(2):258-268

[65] Ahmad AA, Zawawi NA, Kasim FH, Inayat A, Khasri A. Assessing the gasification performance of biomass: A review on biomass gasification process conditions, optimization and economic evaluation. Renewable and Sustainable Energy Reviews. 2016;**53**: 1333-1347

[66] Bilgen S, Kaygusuz K, Sari A. Renewable energy for a clean and sustainable future. Energy Sources. 2004;**26**(12):1119-1129

[67] Saini JK, Saini R, Tewari L. Lignocellulosic agriculture wastes as biomass feedstocks for second-generation bioethanol production: Concepts and recent developments. Biotech. 2014;**3**

[68] Najafi G, Ghobadian B, Tavakoli T, Yusaf T. Potential of bioethanol production from agricultural wastes in Iran. Renewable and Sustainable Energy Reviews. 2009;**13**:1418-1427

[69] Kabel MA, Bos G, Zeevalking J, Voragen AGJ, Schols HA. Effect of pretreatment severity on xylan solubility and enzymatic breakdown of the . remaining cellulose from wheat straw. Bioresource Technology. 2007;**98**(10): 2034-2042

[70] Mushimiyimana I, Tallapragada P. Bioethanol production from agro wastes by acid hydrolysis and fermentation process. Journal of Scientific and Industrial Research. 2016;**75**:383-388

[71] Ingale S, Joshi SJ, Gupte A. Production of bioethanol using agricultural waste: Banana pseudo stem. Brazilian Journal of Microbiology. 2014; **45**(3):885-892

[72] Maiti S, Sarma SJ, Brar SK, Le-Bihan Y, Drogui P, Buelna G, Verma M. Agro-industrial wastes as feed stock for sustainable bio-production of butanol by Clostridium beijerinckii. Food Bioproducts Processing. 2016;**98**: 217-226. DOI: 10.1016/j.fbp.2016.01.002

[73] Kim S, Dale BE. Global potential bioethanol production from wasted crops and crop residues. Biomass and Bioenergy. 2004;**26**(4):361-375

[74] Rashad MM, Mahmoud AE, Ali MM, Nooman MU, Al-Kashef AS. Antioxidant and anticancer agents produced from pineapple waste by solid state fermentation. International Journal of Toxicology and Pharmacology Research. 2015;**7**(6):287-296

[75] Hegazy AE, Ibrahium MI. Antioxidant activities of orange peel extracts. World Applied Sciences Journal. 2012;**18**:684-688

[76] Parashar S, Sharma H, Garg M. Antimicrobial and antioxidant activities of fruits and vegetable peels: A review. Journal of Pharmcogenetics Phytochemistry. 2014;**3**:160-164

[77] Yang SS, Swei WJ. Cultural condition and oxytetracycline production by Streptomyces rimosus in solid state

fermentation of corn cob. World Journal of Microbiology and Biotechnology. 1996;**12**:43-46

[78] Kalogeris E, Christakopoulos P, Katapodis P, Alexiou A, Vlachou S, Kekos D, et al. Production and characterization of cellulolytic enzymes from the thermophilic fungus. Thermoascus aurantiacus under solid state cultivation of agricultural wastes. Process Biochemistry. 2003;**38**: 1099-1104

[79] Beuchat LR. Oncom (fermented peanut presscake). In: Reddy NR, Pierson MD, Salunkhe DK, editors. Legume-based Fermented Foods. Boca Raton: CRC Press; 1986. pp. 135-144

[80] Mak ML. Tempeh production from soyabean milk wastes and soybeans by various Rhizopus strains. B.Sc. Hons thesis, Singapore: Singapore National University; 1986

[81] Muzamil M, Mani I, De D, Lande S. Influence of operational parameters of windrow turner for mass production of compost. International Journal of Agricultural and Statistical Sciences. 2013;**9**(**1**):19-28

Chapter 4

Nutrient Supplying Potential of Crop Residues in Indian Agriculture

Arjun Singh, Saroj Choudhary, Rajendra Prasad Meena and Anchal Dass

Abstract

The Indian agriculture sector has seen tremendous changes over the period. The country's food grain production has increased from just 50 million tons (mt) in the 1950s to 308 mt at present, around a sixfold rise. All of this could become possible due to the introduction of green revolution technologies in the 1970s and onwards, such as the introduction of high-yielding varieties, chemical fertilizers, pesticides, selective herbicides, and improvement of irrigation facilities with improved agronomic management. According to an estimate, chemical fertilizers alone account for 50% of growth in the country's foodgrain production. In the early 1950s, total fertilizer consumption in India was just 0.069 mt, which has increased to 2.25 mt in 1970–71 and 32.5 mt at present. This about 14.4 times rise in fertilizer consumption since the 1970s level to the present highlights the role of fertilizers and nutrient management in Indian agriculture. However, a continuous increase in fertilizer consumption over the year has invited many secondary agricultural problems, such as multi-nutrient deficiency, increased cost of production, and declining factor productivity of fertilizers. Imbalance and overuse use of fertilizers, along with increased cropping intensity and reduced recycling of residues are some of the reasons for such problems. Increased agricultural production also generates a large volume of surplus residue, which often creates problems for farmers. It is well-understood facts from various research studies that crop residues are the hidden treasurer of all the essential nutrients and organic carbon. This chapter highlights the potential of crop residue in nutrient recycling in India and the availability of surplus crop residue.

Keywords: crop residue, nutrient recycling, soil health, nitrogen, phosphorus, potassium

1. Introduction

The agriculture sector alone engages about 55% of the Indian population and is the backbone of the Indian economy. India's foodgrain production has significantly increased from 50 mt in the 1950s to 308.6 mt in 2020, making it the world's second-largest producer [1]. Green revolution technologies (GRTs) and modern infrastructure helped the country achieve self-sufficiency by implementing a variety of modern methods and technologies, including the use of high-yielding varieties, chemical fertilizers, irrigation facilities, pesticides, and farm machinery, all backed by increased public investment and facilitating institutions [2]. The adoption of high-yielding varieties of rice and wheat coupled with chemical fertilizers and agronomic management were critical factors in achieving foodgrain self-sufficiency. Better nutrient management through fertilizers alone accounts for a 50% rise in food grain production in the country [3]. Consumption of fertilizers increased to 32.5 million tons in 2020–21, scaled up from 0.069 mt in 1950–51 and 2.25 mt in 1970–71. India's total NPK consumption increased 471 times to its 1950–1951 level, and 14.44 times to the 1970–1971 level [1–4]. Increased use of chemical fertilizer and intensive agriculture create secondary agricultural problems. More than half of Indian soils are deficient in NPK, indicating a multi-nutrient deficiency. According to a recently published report based on soil samples collected between 2011–2017, about 36.5, 23.4, 12.8, 7.1, and 4.2% of soils were found deficient in Zn, B, Fe, Mn, and Cu, respectively [5, 6]. Multi-nutrient deficiency is of prime concern to Indian agriculture, mostly caused by the over and imbalance usage of primary major nutrients, particularly nitrogen, as well as a lack of attention to secondary and micronutrients, and increased intensive cropping without returning crop biomass. Approximately 188.4 thousand tons (Tt) of micronutrients were removed by 263 mt of foodgrains produced [7]. Based on annual crop production data from 2011–12 to 2015–16, India is producing 696.38 million tons of gross crop residue annually [8]. This hulk size of crop residue warrants judicious management in agriculture, which would aid in the elimination of multi-nutrient deficiency, as well as reduced fertilizers costs and improved soil health. This chapter will touch on the issues of nutrient deficiency in Indian soils and the potential of nutrient recycling from agricultural crop waste.

2. Status of nutrient deficiency in Indian soils

Indian soils are largely poor in fertility status, especially in nitrogen, phosphorus, and potassium, which is compounded by the emergence of micronutrient deficiencies, as a result of their accelerated removal under intensive agriculture. Multi-nutrient deficiency has now become a severe problem in Indian soils; nearly all of the 17 essential plant nutrients are deficient in Indian soils to variable degrees. Multi-nutrient deficiency is exaggerated with modern highly intensive agriculture, which lacks in providing micronutrients back to the soil. According to the latest data from the DAC's soil health dashboard (available at https:/soilhealth .dac.gov.in /NewHome Page/StateWiseNPKChart), 96% of Indian soils have extremely low, and or medium nitrogen content, with only 4% having a high nitrogen content (**Figure 1**).

About 61% of Indian soils are deficient in available phosphorus. Similarly, about 5, 10, and 47% of soils are very low, and medium in plant-available potassium, respectively.

In a recent report from Shukla *et al.*, [5], a detailed description of secondary (sulfur) and micronutrient deficiencies in Indian soils are given in **Figures 2** and **3**.

About, 40.5, 36.5, 12.8, 4.2, 7.1, and 23.2% of soils in 536 districts in India are found to be deficient in sulfur (S), zinc (Zn), iron (Fe), copper (Cu), manganese (Mn), and boron (B), respectively [5]. In Indian soils, zinc was reported to be the most deficient micronutrient. The prevalence of zinc deficiency varies by state, ranging from 9.6% in Uttarakhand to 75.3% in Rajasthan. Zn deficiency was found to

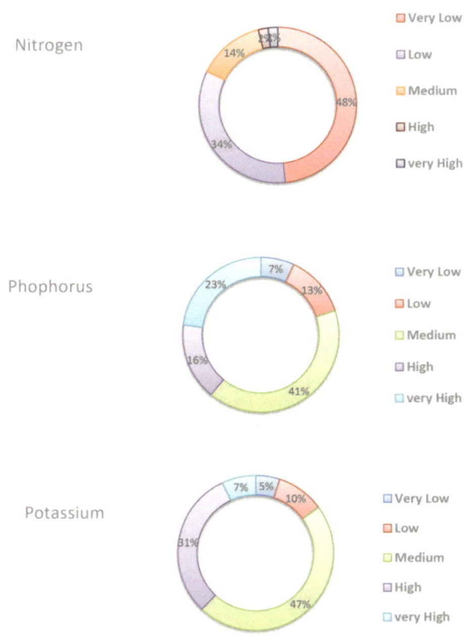

Figure 1.
Primary nutrient deficiencies (%) in Indian soils. (Source: https://www.soilhealth.dac.gov.in/NewHomePage/StateWiseNPKChart).

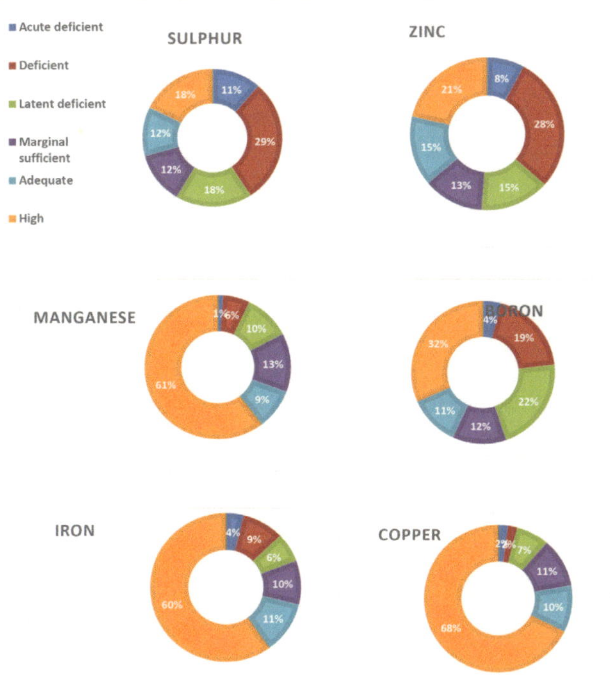

Figure 2.
Percentage of Indian soils deficient in S and micronutrients (adopted from [5]).

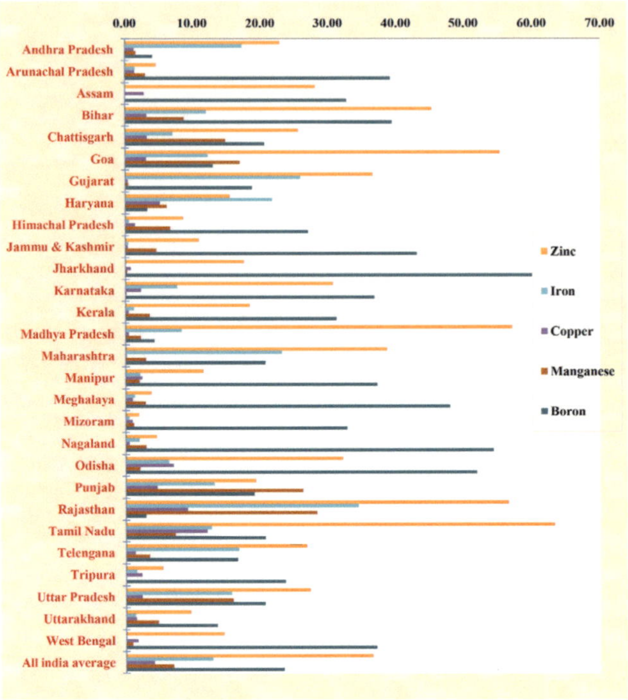

Figure 3.
State-wise percent distribution of micronutrient deficiencies in India (source: [5]).

be more prevalent in the states of Rajasthan (75.3%), Madhya Pradesh (66.9%), Tamil Nadu (65.5%), Maharashtra (54.0%), Bihar (44.0%), and Uttar Pradesh (33.1%); in the rest of the country, deficiency varied from 9.6 to 25%. Boron is the second most deficient (23.2%) micronutrient in Indian soils, trailing just Zn (36.5%). B deficiency is more common in calcareous soils of Bihar (39.39%) and Gujarat (18.72%), as well as acid soils of Jharkhand (60%), Nagaland (54.31%), Odisha (51.88%), Meghalaya (47.93%), West Bengal (37%) and Jammu and Kashmir (43%). The primary cause is excessive B leaching in sandy loam soils, and alluvial and less deposits [9–11]. Iron deficiency is more prevalent in western states, such as Rajasthan (38.34%), Gujarat (25.87%), Maharashtra (23.12%), Haryana (21.72%), and Punjab (21.72%) (13%). Fe deficiency is emerging in states of Andhra Pradesh (12.24%), Telangana (16.65%), Uttar Pradesh (15.56%), Tamil Nadu (12.62)%, and Bihar (12%). In comparison to other micronutrients, manganese deficiency in Indian soils is low (7.10%). Rajasthan (28.28%), Punjab (26.20%), Goa (16.91%), Uttar Pradesh (15.82%), and Chhattisgarh (14.77%) are the states with the highest levels of Mn deficiency. The copper deficiency was observed to be lower in Indian soils.

3. Partial factor productivity: foodgrain production *vis-à-vis* fertilizers consumption

Fertilizer consumption in India surged after the adoption of fertilizer-responsive high-yielding rice and wheat varieties in 1965–66, later termed the green revolution era.

DOI: http://dx.doi.org/10.5772/intechopen.108970

Years	Foodgrains production (mt)	N use (000 tons)	P_2O_5 use (000 tons)	K_2O use (000 tons)	Total ($N+P_2O_5+K_2O$) (000 tons)	N/K_2O ratio	P_2O_5/K_2O ratio	$N+P_2O_5+K_2O$ ratio	PFP (kg/kg)
1950–51	50.8	55	8.8	6	69.8	9.17	1.47	9.17:1.47:1	728.08
1960–61	82.0	211.7	53.1	29	293.8	7.30	1.83	7.30:1.83:1	279.17
1970–71	108.4	1479.3	541	236.3	2256.6	6.26	2.29	6.26:2.29:1	48.05
1980–81	129.6	3678.1	1213.6	623.9	5515.6	5.90	1.95	5.90:1.95:1	23.50
1990–91	176.4	7997.2	3221	1328	12546.2	6.02	2.43	6.02:2.43:1	14.06
2000–01	196.8	11310.2	4382.4	1667.1	17359.7	6.78	2.63	6.78:2.63:1	11.34
2010–11	244.5	16558.2	8049.7	3514.3	28122.2	4.71	2.29	4.71:2.29:1	8.69
2011–12	259.3	17300.3	7914.3	2575.5	27790	6.72	3.07	6.72:3.07:1	9.33
2012–13	257.1	16820.9	6653.4	2061.8	25536.2	8.16	3.23	8.16:3.23:1	10.07
2013–14	265.1	16750.1	5633.5	2098.9	24482.4	7.98	2.68	7.98:2.68:1	10.83
2014–15	252.0	16949.6	6098.9	2532.9	25581.3	6.69	2.41	6.69:2.41:1	9.85
2015–16	251.5	17372.3	6978.8	2401.5	26752.6	7.23	2.91	7.23:2.91:1	9.40
2016–17	275.1	16735.9	6705.5	2508.5	25949.9	6.67	2.67	6.67:2.67:1	10.60
2017–18	285.0	16959.3	6854.4	2779.7	26593.4	6.10	2.47	6.10:2.47:1	10.72
2018–19	284.9	17637.8	6910.2	2680.3	27228.2	6.58	2.58	6.58:2.58:1	10.47
2019–20	297.5	19101.3	7662	2607	29370.4	7.33	2.94	7.33:2.94:1	10.13
2020–21	308.6	20404	8977.9	3153.7	32535.6	6.47	2.85	6.47:2.85:1	9.48

Table 1.
Foodgrain production, NPK consumption, their ratio, and partial factor productivity in India from 1950–51 to 2020–21.

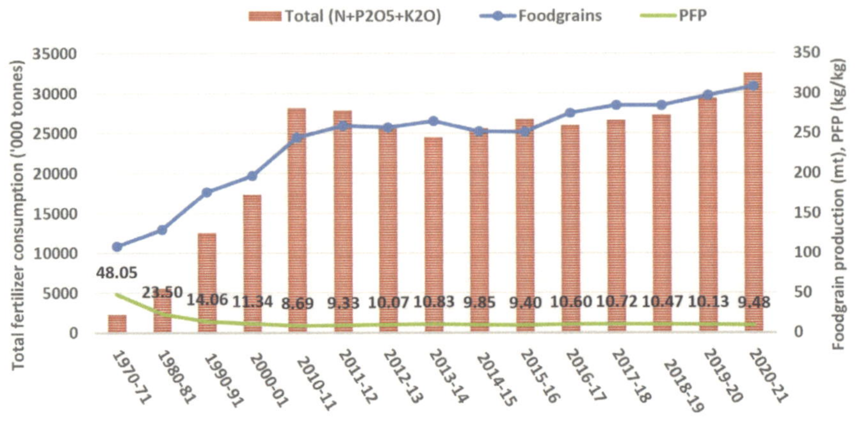

Figure 4.
Fertilizer consumption, foodgrain production, and partial factor productivity.

Consumption of nutrients increased to 32.5 million tons in 2020–21, up from 0.069 Mt in 1950–51 and 2.25 Mt in 1970–71 (**Table 1**, **Figure 4**). With 32.5 mt of total fertilizer nutrient use (N+P2O5+K2O), India is second only to China (52.50 mt) globally. Total nutrient consumption (NPK) increased from 2.25 mt in 1970–71 to 32.5 mt in 2020–21, while foodgrain production rose from 108.4 mt to 308.6 mt during the same period. From 1950 to 1951, India's total NPK consumption increased 471 times, from 1970 to 1971, the increase was 14.44 times. However, partial factor productivity (kg of foodgrain produced per unit of fertilizer nutrient used) fell from 728 kg/kg in 1950–51 to 9.48 kg/kg in 2020–21 (**Figure 1**).

The decline in PFP at an alarming rate is a sign of decreasing crop response to fertilizer use and declining soil health. The relative use of nitrogen, phosphorus, and potash fertilizers remains highly skewed toward N. In 1950–51, the NPK use ratio was 9.17:1.47:1, which decreased to 6.26:229:1 in 1970–71, and 6.47:285:1 in 2020–21 (**Table 1**, **Figure 4**). Use of micronutrients remains negligible in Indian agriculture, which is evident from emerging multi-nutrient deficiencies (**Table 2**).

Crop	N (%)	P (%)	K (%)	References
Rice	0.45–0.49	0.089–0.104	1.45–1.51	[12]
	0.55–0.66	0.067–0.103	1.463–1.525	[13]
	0.61	0.18	1.38	[14]
	0.71	-	-	[15]
Wheat	0.37	-	1.2	[16]
	0.34	-	-	[15]
	0.62	0.038	1.324	[17]
	0.48	0.16	1.18	[14]
Maize	0.6		--	[15]
	0.52	0.18	1.35	[14]

Crop	N (%)	P (%)	K (%)	References
Sorghum	0.52	0.23	1.34	[14]
	0.537	0.194	1.59	[18]
Pearlmillet	0.45	0.16	1.14	[14]
	0.68	0.194	1.39	[19]
Pulses	1.29	0.36	1.64	[14]
Gram	0.6	0.223	-	[20]
Redgram	0.74	0.28	0.89	[21]
Lentil	1.96	0.199	1.142	[13]
Oilseed	0.8	0.21	0.93	[14]
Groundnut	1.6	0.23	1.37	[14]
Soybean	0.85	-	1.45	[16]
	0.58	-	-	[15]
	1.93	0.14	1.84	[22]
Mustard	0.55–0.65	0.25–0.32	0.39–0.41	[23]
Sunflower	0.55–0.58	0.18–0.22	1.70–1.79	[24]
Sugarcane	0.4	0.18	1.28	[14]
Cotton	0.67	-	0.46	[16]
	0.86	0.3	0.92	[21]
Banana	0.47	0.062	3.84	[25]
	0.7	0.13	7.71	[26]

Table 2.
Major nutrients content in residue of major field crops.

4. Availability of crop residue in India

Crop residues or biomass are non-economical material remains after the removal of economically important parts (generally grains in cereals, seeds in pluses and oilseeds, pods/fruits in vegetables, and fruits) at the harvesting sites and agricultural processing sites [27]. The amount of crop waste is increasing with increasing crop area and cropping intensity. Crop biomass is often treated as wasteful material and is subjected to burning, exporting for other uses, selling, and feeding animals. However, multiple studies have emphasized the importance of crop residues in sustainable agriculture and conservation agriculture, as and means of reducing the chemical fertilizer requirement [16, 28–30].

Venkatramanan *et al.,* [8] calculated the annual gross crop residue potential (GCR) of India, 696.38 million tons, based on the annual crop production data from 2011–12 to 2015–16, which includes cereal crops, pulse crops, oilseed crops, sugarcane, fiber crops and horticultural and plantation crops (**Table 3**, **Figures 5** and **6**). The major contributors to GCR are cereals (52%), bananas and plantain (17.76%), sugarcane (17.13%), oilseed crops (6.26%), fiber crops (4.5%), and pulse crops (1.95%). Cereals contribute about 364.3 mt to the total gross crop residue generated annually in India. The sugarcane crop alone generates about 119 mt/year of gross

Crop	Gross crop residue Potential (mt)	Surplus Crop residue (mt)	Nutrient supply potential (kg nutrients/t residue)			Gross nutrient supply potential (mt)			Nutrient supply potential based on surplus residue (mt)		
			N	P	K	N	P	K	N	P	K
Rice	156.89	46.91	6.00	1.20	14.50	0.94	0.19	2.27	0.28	0.06	0.68
Wheat	149.05	48.67	4.50	1.00	12.30	0.67	0.15	1.83	0.22	0.05	0.60
Maize	43.222	12.83	5.60	1.80	13.50	0.24	0.08	0.58	0.07	0.02	0.17
Sorghum	10.26	3.01	5.30	2.10	14.70	0.05	0.02	0.15	0.02	0.01	0.04
Pearlmillet	4.85	1.52	5.70	1.80	12.70	0.03	0.01	0.06	0.01	0.00	0.02
Gram	7.1	2.7	6.00	2.20	16.00	0.04	0.02	0.11	0.02	0.01	0.04
Redgram	4.93	2.04	7.40	2.80	8.90	0.04	0.01	0.04	0.02	0.01	0.02
Lentil	1.54	0.49	19.60	2.00	11.40	0.03	0.00	0.02	0.01	0.00	0.01
Sugarcane	119.34	44.94	4.00	1.80	12.80	0.48	0.21	1.53	0.18	0.08	0.58
Groundnut	14.46	3.51	16.00	2.30	13.70	0.23	0.03	0.20	0.06	0.01	0.05
Mustard	9.73	2.21	6.00	2.90	4.00	0.06	0.03	0.04	0.01	0.01	0.01
Linseed	0.125	0.031	8.00	2.10	9.30	0.00	0.00	0.00	0.00	0.00	0.00
Safflower	0.305	0.09	8.00	2.10	9.30	0.00	0.00	0.00	0.00	0.00	0.00
Soybean	17.85	4.59	11.20	1.40	16.50	0.20	0.02	0.29	0.05	0.01	0.08
Sunflower	1.1	0.33	5.70	2.00	17.50	0.01	0.00	0.02	0.00	0.00	0.01
Cotton	28.6	12.73	7.70	3.00	6.90	0.22	0.09	0.20	0.10	0.04	0.09
Banana	68.9	27.55	5.90	1.00	57.80	0.41	0.07	3.98	0.16	0.03	1.59
Total	638.25	214.15	132.60	33.50	251.80	3.65	0.94	11.34	1.20	0.32	3.98

Table 3.
Nutrient supplying potential of annually generated crop residue in India (Based on potential and surplus residue, source: [8]).

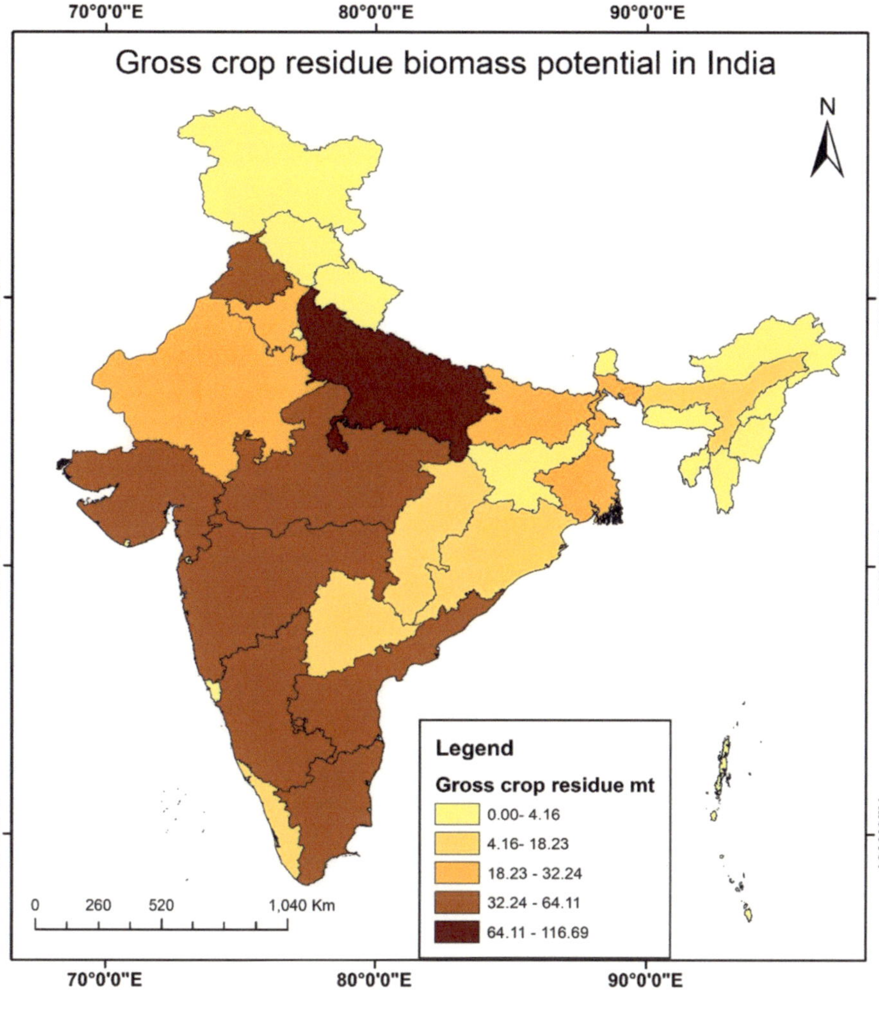

Figure 5.
Gross crop residue potential of India (Adopted from [8]).

crop residue. The contribution of oilseeds and pulses is only 43.57 million tons/year and 13.58 million tons/year of gross crop residues, respectively. Uttar Pradesh, Maharashtra, Karnataka, Madhya Pradesh, Tamil Nadu, Punjab, Gujarat, and West Bengal all contribute considerably to crop residue generation (**Figures 5** and **6**).

The surplus residue is an unutilized part of the gross crop residue, which is not being used by farmers, and it can be exploited for nutrient recycling in the agro-ecosystem. Surplus crop residues available for nutrient recycling are estimated to be 214.15 million tons annually, which is nearly 33% of gross crop residue potential. Within the crop category, cereals and sugarcane contribute to an extent of 53.86% and 21.21% to the surplus crop residue in India, respectively.

Rice, wheat, and sugarcane crops all contribute a nearly equal amount to gross residual surplus. About 27.5 million tons of banana crop residue (peels) is added to the estimated surplus crop residue. Fiber crops contribute near about 13.6 million tons in crop residual surplus (stalk, husk, and boll shell). Surplus crop residues from oilseeds and pulse crops amount to 10.8 mt and 5.2 mt, respectively. Uttar Pradesh produces

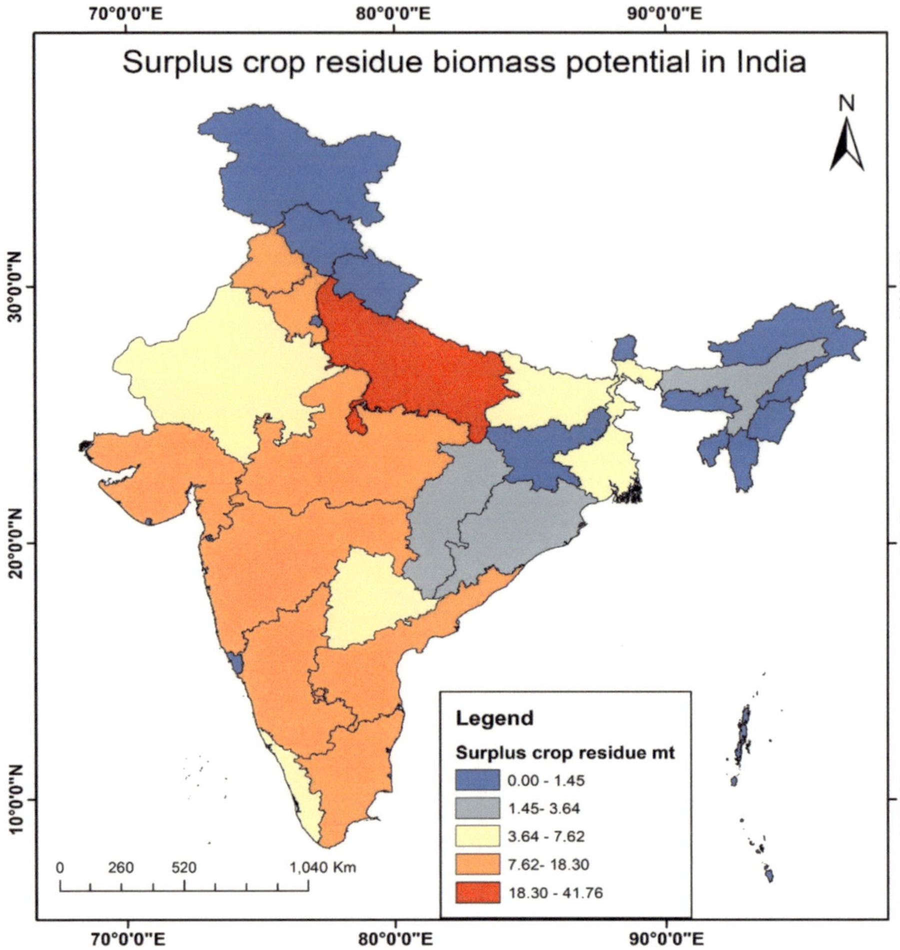

Figure 6.
Surplus crop residue potential of India (Adopted from [8]).

116.7 mt of total crop waste and 41.8 mt of surplus crop residues, according to estimates. States of Punjab, Madhya Pradesh, Gujarat, Maharashtra, Karnataka, Andhra Pradesh, and Tamil Nadu produce between 32.2 mt and 64.1 million tons of gross crop residues (**Figures** 5 and **6**).

5. Nutrient supplying potential of crop residues in India

Crop residue is an important source of essential plant nutrients; therefore, it should be explored for recycling in agriculture. It is widely known that depending on the crop species, varying amounts of N, P, K, and other nutrients are extracted from and returned to the soil *via* crop residues [12–14].

The amount of nutrients that crop residue can provide, as well as the rate at which it decomposes and becomes available to plants, is determined by the type of residue. Based on 17 primary crops, including bananas, India generates 638.25 mt and 214.15 mt of total and surplus agricultural residue, respectively. The potential of nutrient supply from these residues varied with the nutrient content in the residue. Major nutrient content (NPK) in major field crops is given in **Table 2**. Nitrogen concentration in rice residue ranges from 0.45 to 0.71%, phosphorus from 0.089 to 0.18%, and potassium from 1.2 to 1.52%. Nitrogen content is higher in pulses (1.29%) and legume oilseeds like soybean (1.93%) and groundnut (1.6%). Phosphorus content is relatively lower than N and K, It varied from 0.089 to 0.36%. K content in crop biomass was found to have higher variation among the type of crops, being higher in bananas (3.84–7.7%) to as low as 0.89% in gram.

The per ton of rice residue could supply 6, 1.20, and 14.50 kg of N, P, and K, respectively (**Table 2**). Similarly, one ton of lentil residue could supply 19.6 kg N and 11.4 kg K. The variation in the nutrient amount of various crop residues depends on the percentage of nutrient content in the particular crop. The potassium supply from one ton of banana residue could be 57 kg. Based on the gross crop residue generated (638.25 mt annually), the nutrient locked in or present in is 16.02 mt of NPK, comprising 3.65 mt of N, 0.94 mt of P, and 11.34 mt of K. However, with the competition for residues in other sectors, the surplus reside available for recycling in the field is 214.15 mt [8], which would supply a total of 5.5 mt of NPK, comprised of 1.20 mt of N, 0.32 mt of P, and 3.98 mt of K. Similarly, Srivastava [31] reported that ten major crops (rice, wheat, sorghum, pearl millet, barley, finger millet, sugarcane, potato tubers, and pulses) of India generate about 312.5 mt of crop residues that have the nutrient supplying potential of about 6.46 million tons of NPK.

6. Conclusion

Crop residue often being treated as wasteful material and burnt on-farm. Annually, India generates about 214.1 mt surplus crop residue, which could supply 5.5 mt of NPK. However, the policy, strategies, and methods need to be explored for better utilization of the whopping amount of crop residue in India. Currently, crop residues are either burnt on the field or used for ruminants as feed, or transported for other uses. To exploit the trapped nutrients in crop biomass, mass awareness and government-backed schemes have to be implemented. The recycling and reuse of crop biomass would not only reduce the dependence on chemical fertilizers but will also enhance the functioning of soil health *via* increasing various soil physical, chemical, and biological properties.

Author details

Arjun Singh[1,2*], Saroj Choudhary[3], Rajendra Prasad Meena[4] and Anchal Dass[2]

1 ICAR- National Research Centre for Banana, Tiruchirappalli, Tamil Nadu, India

2 ICAR-Indian Agricultural Research Institute, New Delhi, India

3 Department of SSAC, Banaras Hindu University, Varanasi, Uttar Pradesh, India

4 Crop Production Division, ICAR-VPKAS, Almora, India

*Address all correspondence to: arjunsiari@gmail.com

References

[1] Pocket Book of Agricultural Statistics. Directorate of Economics & Statistics Department of Agriculture, Cooperation & Farmers Welfare, Ministry of Agriculture & Farmers Welfare, Government of India. 2020;23.

[2] NAAS. Zero Budget Natural Farming - A Myth or Reality? New Delhi: National Academy of Agricultural Sciences; 2019

[3] Shukla AK, Behera SK, Chaudhari SK, Singh G. Fertilizer use in Indian Agriculture and its impact on human health and environment. Indian Journal of Fertilizers. 2022a;**18**(3):218-237

[4] FAI. Fertilizer Statistics 2019-20. 65th ed. New Delhi: The Fertiliser Association of India; 2020

[5] Shukla AK, Behera SK. All India research project on micro- and secondary nutrients and pollutant elements in soils and plants: Research achievements and future thrusts. Indian Journal of Fertilizers. 2019;**15**:522-543

[6] Shukla AK, Behera SK, Pakhre A, Chaudhari SK. Micronutrients in soils, plants, animals and humans. Indian Journal of Fertilizers. 2018;**14**:30-54

[7] Takkar PN, Shukla AK. Management of soil fertility: Micronutrients. In: Pathak H, Sanyal SK, Takkar PN, editors. State of Indian Agriculture – Soil. New Delhi: NAAS; 2015. pp. 121-152

[8] Venkatramanan V, Shah S, Prasad S, Singh A, Prasad R. Assessment of bioenergy generation potential of agricultural crop residues in India. Circular Economy and Sustainability. 2021;**1**:1335-1348

[9] Shukla AK, Behera SK. Micronutrient fertilizers for higher productivity. Indian Journal of Fertilizers. 2012;**8**(4):100-117

[10] Shukla AK, Tiwari PK. Micro and secondary nutrients and pollutant elements research in India. In: AICRP on Micro and Secondary Nutrients and Pollutant Elements in Soils and Plants, ICAR-IISS. Bhopal; 2016. pp. 1-196

[11] Takkar PN. Micronutrient research and sustainable agricultural productivity in India. Journal of the Indian Society of Soil Science. 1996;**44**:562-581

[12] Singh A, Das A, Singh CV, Dhar S, Sudhishri S, Das K, et al. Growth, productivity and nutrient concentration of aerobic rice (Oryza sativa L.) under different planting methods, irrigation schedules and soil adjuvant application. Annals of Agricultural Research. 2017;**38**(4):368-374

[13] Choudhary S. Thesis: Effect of Sulphur and Boron on yield, nutrients uptake and protein quality of lentil grown in Vindhyan red soil of Mirzapur. Varanasi: BHU; 2021

[14] Tandon HLS. Organic resources an assessment of potential supplies, their contribution in agricultural productivity, and policy issues for Indian Agricultural from 2000-2025. In: Kanwar JS, Katyal JC, editors. Plant Nutrient Needs, Supply Efficiency and Policy Issues: 2000-2025. New Delhi, India: National Academy of Agricultural Sciences; 1997. pp. 15-28

[15] Lenka S, Choudhary R, Lenka NK, Saha JK, Amat D, Patra AK, et al. Nutrient management drives the direction and magnitude of nitrous oxide flux in crop residue-returned soil under

different soil moisture. Frontiers in Environmental Science. 2020;**10**:857233. DOI: 10.3389/fenvs.2022.857233

[16] Hulugalle NR, Weaver TB. Short-term variations in chemical properties of vertisols as affected by amounts, carbon/nitrogen ratio, and nutrient concentration of crop residues. Communications in Soil Science and Plant Analysis. 2005;**36**(11-12):1449-1464

[17] Hirzel J, Undurraga P, León L, Matus I. Effect of residue incorporation on wheat grain and plant nutrients during a four-year rotation of canola-wheat and bean-wheat crops. Communications in Soil Science and Plant Analysis. 2022

[18] Jat MK, Purohit HS, Singh HS, Garhwal RS, Choudhary M. Effect of integrated nutrient management on yield and nutrient uptake in sorghum (Sorghum bicolor). Indian Journal of Agronomy. 2013;**58**(4):543-547

[19] Choudhary M, Rana KS, Kumar P. Nutritive value of pearl millet stover as influenced by tillage, crop residue and sulphur Fertilization. Range Management & Agroforestry. 2019;**40**(1):150-155

[20] Balai K, Jajoria M, Verma R, Deewan P, Bairwa SK. Nutrient content, uptake, quality of chickpea and fertility status of soil as influenced by fertilization of Phosphorus and Zinc. Journal of Pharmacognosy and Phytochemistry. 2017;**6**(1):392-398

[21] Aparna K, Rekha B, Vani KP, Ram PT. Nutrient dynamics of cotton and red gram residue compost. Current Science. 2022;**122**(6):25

[22] Tiwari R, Sharma YM, Dwivedi BS, Mitra NG, Kewat ML. Nutrient content and uptake by soybean as influenced by continuous application of fertilizer and manure in black soil. Journal of Pharmacognosy and Phytochemistry. 2019;**8**(4):140-144

[23] Om H, Rana K, Ansari MA. Productivity and nutrient uptake of mustard (Brassica juncea) influenced by land configuration and residual and directly applied nutrients in mustard under limited moisture conditions. Indian Journal of Agricultural Sciences. 2013;**83**(9):933-938

[24] Babu S, Rana DS, Choudhary AK. Effect of sunflower stover and nutrients management on energetics, nutrient acquisition and soil nutrient balance of pigeonpea (*Cajanus cajan*)–sunflower (*Helianthus annuus*) cropping system. Indian Journal of Agronomy. 2014;**59**(4):549-555

[25] Wortman CS, Karamura EB, Gold CS. Nutrient flows from harvested Banana Pseudostems. African Crop Science Journal. 1994;**2**(2):179-182

[26] Ultra VC Jr, Mendoza DM, Briones AM. Chemical changes under aerobic composting and nutrient-supplying potential of banana residue compost. Renewable Agriculture and Food Systems. 2005;**20**(2):113-125

[27] Prasad S, Singh A, Korres N, Rathore D, Sevda S, Pant D. Sustainable utilization of crop residues for energy generation: a life cycle assessment (LCA) perspective. Bioresource Technology. 2020;**303**:122964

[28] Mahajan A, Gupta RD. Potential of organic resources as plant nutrients in India. In: Mahajan A, Gupta RD, editors. Integrated Nutrient Management (INM) in a Sustainable Rice—Wheat Cropping System. Dordrecht: Springer; 2009

[29] Shyamsundar P, Springer N, Tallis H, Polasky S, Jat M, Sidhu H, et al. Fields on fire: Alternatives to crop residue burning in India. Science. 2019;**365**:536-538

[30] Singh B, Singh Y. Management of crop residues in the Indo-Gangetic plains. In: Singh Y, Singh B, Nayyar VK, Singh J, editors. Integrated Plant Nutrient Supply Systems for Sustainable Rice-wheat Rotation. Ludhiana, Punjab, India: National Agricultural Technology Project, India and Punjab Agricultural University; 2003. pp. 286-301

[31] Srivastava S. Potential and scope of organic resources toward nutrient supply in Indian Agriculture IN: Manual of ICAR Short Course on "Concepts and Mechanisms of soil carbon sequestration and stabilization for soil health improvement and climate change mitigation" during 02-11March 2022 at ICAR-IISS, Bhopal, M.P., India. 2022; p. 33

Chapter 5

Agro-Industrial Waste Management: The Circular and Bioeconomic Perspective

Cosmas Chikezie Ogbu and Stephen Nnaemeka Okey

Abstract

Traditional agricultural production is circular. Virtually no waste is produced. Residues are returned to soil as compost; used as bedding material in livestock husbandry (and returned to soil as compost) or as feed to produce animal protein and manure; utilized as construction materials; or fuel for domestic energy. Circular agricultural production ensures soil conservation, waste reduction, residues reuse, and recycling. The ever rising global population, and demand for food and agro-industrial products, necessitated a transition to linear agricultural production which generates enormous quantities of agricultural residues, agro-industrial, and food wastes. The economic losses, environmental degradation, and health hazards resulting from poor management of excess wastes, and their mitigation have been the subject of research and policy efforts at continental and regional levels. Current waste management models redirect attention to circular agricultural production and bioeconomic approaches aimed at waste reduction, reuse, and recycling. Such approaches view agricultural wastes as raw materials with economic benefits for the farmer, consumer, and investor in varied industrial enterprises (crop and animal production, animal and human health, food, beverage, neutraceutical, pharmaceutical, cosmetics, and material industries). The present review attempts to collate information on global production, and possible valorization of recyclable agro-industrial residues and food wastes.

Keywords: wastes, theoretical and technical availability, valorization, bioactive compounds, value-added products

1. Introduction

Agro-industrial wastes are inedible materials produced as a result of various agricultural and agro-industrial operations. They include wastes from slaughter-houses and meat processing, animal dung or manure, field crop wastes, crop residues, harvest wastes, and wastes from food consumption and processing [1, 2]. The huge diversity characteristic of the agricultural and livestock sectors means that very large and heterogeneous products end up as wastes. There are hence several types of agro-industrial wastes based on material composition and management.

1.1 Types of agro-industrial wastes

Agro-industrial wastes can be divided into three broad categories, namely recyclable and compostable or naturally occurring agricultural and agro-industrial wastes, non-recyclable and non-compostable agricultural and agro-industrial wastes, and hazardous agricultural and agro-industrial wastes. Compostable wastes are recyclable wastes, which can be reused in the farm or recycled in recycling plants. Some such as pruning, straw, leaves, stover, stalk, bagasse, cob, and animal dung or manure are regarded as primary residues because they arise directly from crop and animal production activities while others such as pit, shell, peels, husk, cake, slurry, and slaughterhouse wastes are regarded as secondary wastes because they arise from agro-allied industrial processing [1]. Generally, primary and secondary residues are categorized as least problematic in management. Non-recyclable agro-industrial wastes are wastes that result from farm construction operations, farm mechanization, transport, and livestock protection facilities. They are the most problematic to manage since they are usually bulky and not reused or recycled on-farm. They include plastic sheets and containers, metal containers and equipment, tires, shadings or anti-stone nests, machinery, metal structures for fences or covers, and irrigation facilities. Hazardous agro-industrial wastes are wastes that pose very serious immediate and remote problems if not correctly managed. They include phytosanitary products, chemical containers, acids, fertilizers, waste water, chemical contaminated water, foods, and other materials; medicines, agro-chemicals, and detergents. These wastes are managed following laid down regulations from the appropriate authorities. In this review we focus on the utilization of recyclable (primary and secondary) agro-industrial wastes (agro-industrial residue) (**Figure 1**) for the production of renewable energy and functional products for household, environmental, industrial, medical, veterinary, and animal production applications.

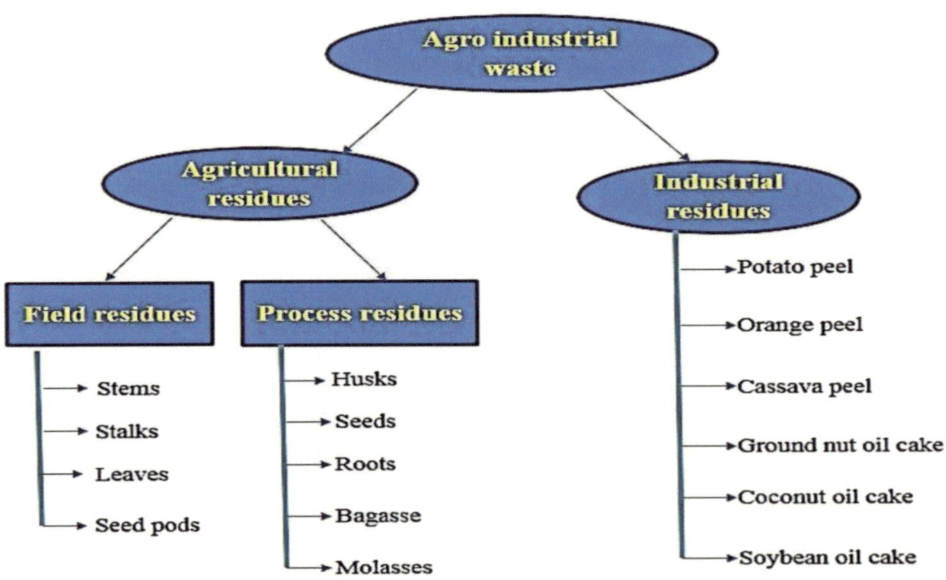

Figure 1.
Classification of recyclable agro-industrial wastes. Source: Sadh et al. [3].

2. Global availability and estimates of agro-industrial residue biomass

Agriculture-based industries produce vast amounts of recyclable residues and wastes [3]. Food production, transportation, storage, processing, distribution, and consumption yield enormous wastes such as crop residues, food wastes, animal manure, animal wastes, and by-products, and forestry waste biomass. For instance, juice industries produce huge amounts of wastes such as peels, pulp, and drupes [4]; beverage industries produce cocoa, and coffee pulp, pod, and stalk as wastes; processing of cereals, canes, and grains yields husks, bran, bagasse, and molasses; nuts yield shells, cakes, and slurry following oil extraction; palm fruit processing yields large quantities of empty palm fruit bunch, palm fruit fiber, and palm oil sludge; processing of palm kernel yield shells while extraction of palm kernel oil yields palm kernel cake; meat industries produce trimmings, bone, offal, feather, hair or fur, cartilage, and blood [1–3]. From the growth of crops, various kinds of primary residues and wastes, namely stalk, stover, straw, leaves, stem, bagasse, and cobs are produced [1, 3]. Raising of livestock and poultry in large confined feeding operations produces nearly unmanageable concentrations of dung or manure and slurry in addition to various slaughterhouse wastes [1, 2].

Global agricultural residue potential is rendered in terms of technical, theoretical, economic, and sustainable potentials [5, 6]. The theoretic potential gives the gross quantity of biomass potentially produced [7] while the technical potential is the fraction of the theoretical potential technically recoverable allowing for economic, social, environmental, and political constraints [6, 8]. The economic potential indicates the fraction of the theoretical and technical potential available for purchase as source of revenue to the producer. Globally, approximately 147.2 million metric tons (Mt) of fiber sources are theoretically available while 709.2 and 673.3 Mt of wheat straw and rice straw residues were estimated, respectively, in 1994 [3, 9]. Residues from cereals and sugar cane production account for 80% of the total residue from crops and constitute the most harvestable biomass [1, 10, 11]. Cho et al. [12] estimated global annual rice straw, wheat straw, corn straw, sugarcane bagasse, and rice husk production at 731, 354, 204, 181, 110 Mt, respectively, while wood biomass waste was put at 4.6 Gt yr^{-1}. Wastes from coffee and olive oil industries were estimated at 7.4 and 30 Mt yr^{-1}, respectively. Capanoglu et al. [13] indicated that close to 1.3 billion tons of food (about one-third of food produced) is lost as wastes before or after reaching the consumer.

Country-specific estimates of agro-industrial and food waste biomass production and availability are scanty and very patchy since very few countries consistently track (document) residue and waste production and use [1]. Thus, reported statistics are results of modeling studies mostly based on national crop production, crop yield, residue-to-product ratio (RPR), area under cultivation, and moisture content [7]. Reports from a number of such studies estimate on a global level an appropriation (including for energy) of 2.9 billion tons yr^{-1} (66% of total annual production) [1, 14, 15]. A 2006–2008 estimate of crop residue production from barley, maize, rice, soybean, sugar cane, and wheat gave 3.7 + 1.3 or – 1.0 Pg (billion tons) accounting for ¾ of total production [16]. Regions outstanding include North and South America, Eastern and Southern Asia, with a production of more than 500 Tg (Mt) yr^{-1} each. South-east Asia and Eastern Europe have estimated value of 200 Tg yr^{-1} each [16]. Earlier estimates include 3.5–4.0 Pg yr^{-1} in the 1990s with cereals, sugar crops, and oil crops accounting for 79% of the total [17], 3.4 Pg yr^{-1} in 1991, and 3.8 Pg yr^{-1} in 2001 with cereals accounting for 74–75% [10], 4.4 Pg yr^{-1} in 2000 [14], and 5.4 Pg yr^{-1} for 1997–2006 using crop-specific harvest indices as estimator instead of RPR [18].

Using agricultural production data from FAO [19], Cooper and Laing [20] quantified crop residues and animal wastes produced on the African continent. The authors indicated that crop residues were mainly from coconut, maize, rice, and sugarcane production. Overall, 639,600 tons of coconut husks and 191,880 tons of coconut shells were estimated. Major producers were Tanzania (140,000 and 42,000 tons, respectively), Ghana (122,000 and 36,600 tons, respectively), and Mozambique (120,000 and 36,000 tons, respectively). Residues from maize were 16,296,301 tons of cobs and 90,602,879 tons of stalks with major production from South Africa (SA) (3,981,199, and 22,134,353 tons, respectively), Egypt (2,405,371 and 13,373,188 tons, respectively), and Nigeria (2,059,759, and 11,451,685 tons, respectively). A total of 22,858,042 tons of sugar cane residues were estimated with SA (6,302,133 tons), Egypt (4,112,925 tons), and Mauritius (1,443,750 tons) leading other countries in the continent. For the USA, the "Billion ton annual supply study" [21] and its update [22] reported an annual crop residue production of 550 Million dry ton matter(Mdt) yr^{-1}; a more recent study reported 518 Mdt yr^{-1} [23] with 5.6 Mt (1% of total) corn stover appropriated for energy production [16, 23]. The US DOE [24] project estimated that approximately 144 Mt of primary agricultural residues are in use in diverse applications across the United States made up majorly of corn stover and concentrated in the Midwest regions, including the states of Illinois, Indiana, Iowa, Kansas, Minnesota, Missouri, Nebraska, Ohio, and South Dakota. Kim and Dale [25] had reported that corn production yielded roughly 10 Mt of grain ha^{-1} and approximately the same amount of stover (assuming a 1:1 crop:residue ratio). An estimate of US biomass resources in terms of economic potentials gave 94 Mt at \$66 ton^{-1} coming from barley straw, corn stover, oats straw, sorghum stubble, and wheat straw with an increase between 158 and 180 Mt by 2040 [1, 24]. For Canada, about 48 Mt of dry agricultural crop residue is possible baring such factors as crop yield, cost of production, collection and transport, properties of residues, distance to processing facilities, the profit potentials of targeted use, and the degree of substitutability of other feedstock [1]. Bloomberg [26] had projected the potential supply of agricultural residues (80% grain straw) in the EU to be approximately 170 Mt at an average supply cost of €67 ton^{-1} while de Wit and Faaij [27] projected approximately 200 Mt at plant gate cost of €51 ton^{-1}. Generally, the cost of crop production, crop yield, cost of residue harvesting, and the supply chain from point of collection to point of processing significantly impact price of residue. For Denmark, Energinet.dk [28] projected 1.0–1.5 Mt yr^{-1} while annual production of crop residues in Canada over the period 2001–2010 was estimated at 82 Mdt [29]. Ji [30] assessed the potential lignocellulosic biomass or crop residues feedstock in China for biofuel production and found a theoretical amount of 930.8 Mt. In Iran, Alavijeh and Yaghmaei [31] reported 11.33 Mt. Pradhan and Mbowa [32] had stated that availability of reliable feedstock data for biofuel production in SA is a challenge. Barahira et al. [33], however, estimated an annual crop residue production of ~43 Mt in SA with ~32 Mt from grains and ~6 Mt from sugar cane. Other estimates include residue from oil crops (groundnut, soybean, sunflower) (~3 Mt), vegetable crops (potato, tomato, cabbage) (~1 Mt), and other minor crops (~0.8 Mt). Residue from maize ranked first among all crops with ~28 Mt followed by sugar cane, wheat (~3 Mt), sunflower (~2 Mt), and soybean (~1 Mt) [33]. Batridzirai et al. [7] had reported the gross (above ground) crop residue potential from maize and wheat in SA as 14.4 Mt yr^{-1}, but only 6.0 Mt yr^{-1} can be removed sustainably from the field. Regions in SA with the highest potentials for residue production include Northern Cape, Mpumalanga, and Free-state accounting for 87% of national residue potential. Maize stover is the predominant crop residue accounting for 90% of the current and future total residue potential. In India, about 500–550 Mt of agro-industrial residues are generated per year [34–36]. Agriculture alone generates 140 Mt yr^{-1} of biomass [34]. Cereals lead (352 Mt, 70%), fiber (66 Mt, 13%), oil seeds (29 Mt), pulses (13 Mt), and sugar cane (12 Mt, 2%). Of the 70% by cereals, rice has 34%, wheat 22% [37]. For fiber, cotton leads (53 Mt, 11% of crop residues) followed by coconut (12 Mt) [37]. FAO [38] reported an estimated total crop residue production of 95 Mt yr^{-1} for member states of the Union économique et monétaireouest-africaine (West African Economic and Monetary Union) (Benin, Burkina Faso, Côte d'Ivoire, Guinea-Bissau, Mali, Niger, Senegal and Togo) in 2010 with cereal straw accounting for 80 Mt yr^{-1} and straw from millet and sorghum ranking highest. Sahelian countries (Niger, Burkina Faso, Mali, and Senegal) had the largest share (96%) of crop residues especially cereal residues. Residues (peels) from tuber crops (yam and cassava) were estimated at 6 Mt yr^{-1} and came mostly from sub-humid countries (Cote d' Ivoire and Southern part of Benin). Agro-industrial by-products include cotton cake (2 Mt yr^{-1} in 2005 but 1.34 Mt yr^{-1} in 2009), groundnut cake (2.5 Mt yr^{-1} in 2009), cereal bran (millet: 1.5–1.8 Mt yr^{-1}; sorghum: 1.3 Mt yr^{-1}), and molasses (100,000 tons yr^{-1}) with Cote d' Ivoire and Senegal accounting for 45 and 15% of the total, respectively. In Nigeria, agro-industrial residues result from crop and animal farming, crop and animal processing, and forestry and timber production [39]. Major crop residues are from cassava, yam, potatoes, fruits and vegetables, plantain, cocoa, coconut, coffee, cowpea, groundnut, maize, millet, rice, sorghum, sugar cane, wheat, soybean, and oil palm [39, 40]. Agba et al. [41] reported that the biomass potential of Nigeria as at 2005 stood at 13 million hectares of fuel wood, 61 Mt yr^{-1} of animal waste, and 83 Mt yr^{-1} of crop residues. From 10 crops, Simonyan and Fasina [42] estimated crop residue availability of 145.6 Mt yr^{-1} while Isola et al. [43] and Okeh et al. [44] reported 227,500 tons day^{-1} of animal manure. The annual production of agricultural wastes is this high because about 94% and 68% of households are engaged in crop and livestock farming, respectively [45]. In a study that evaluated global production of endocarp tissue from horticultural fruit crops, particularly drupes, as residue for biofuel production, Mendu et al. [4] reported an estimate of 2.4×10^7 tons consistent with FAO's 3.1×10^7 tons (29% variation). The study reported greatest density of drupe endocarp production in developing countries in South Asia and broad lower-density distribution across Southern and Northern Europe and the Middle East. Isolated productions were mapped to USA, Africa, China, Australia, Central America, and South America. Highest endocarp yield was from coconut and mango (1.31×10^7 and 3.99×10^6 tons, respectively), which together accounted for 72% of total global drupe endocarp production [4]. Smil [17] had indicated that over 60% of global crop residues are produced in low-income countries, and almost 45% of residues come from the tropics. This is despite considerable regional differences in fraction of residue harvested and used (29% in sub-Saharan Africa and 90% in Western Europe) [14].

3. Management of agro-industrial residues: the circular agricultural and bioeconomy perspectives

Traditional (subsistence) agriculture is based on circular sustainability model, which ensures practically nil waste as residues are recycled or used for various purposes including maintenance of soil fertility [46, 47]. Rise in global population, however, necessitated the intensification of agricultural production, linear-agricultural production system, globalization of food distribution, extensive storage, and agro-industrial processing, all of which generate extensive quantities and varieties of agro-industrial and food wastes [47, 48]. Previously considered of little or no economic value, harvest leftovers, crop residues, animal wastes (bones, carcasses, blood, fat, feathers, hair, cartilages, skins, viscera, and dung), and food wastes (household, food service, and retail wastes) are today viewed as valuable resources of significant economic value such that they have become co-products or raw materials from agro-industrial processing, crop, forestry, and animal production [49]. In the traditional circular agricultural model, farmer's decision on agricultural residue use reflects their needs and preferences [50], and access to and affordability of alternative biomass resources determine the opportunity costs for a farmer or household to sell, use, or replace residues [50]. From time, agricultural waste biomass has fulfilled vital roles including livestock feed, animal bedding material, domestic fuel, construction material, some cash through sales, and maintenance of soil fertility [50–52]. Today, the huge excess residues and wastes after fulfilling these traditional roles are disposed by burning; a practice viewed by farmers as the most convenient, cheap with regard to time, labor, and finance, and beneficial for control of weeds, crop pests, and diseases (**Figure 2**) [52–54]. The traditional agricultural residue management models, however, have poor economic returns to the farmer, and poor soil health conservation and maintenance [52]. In addition, burning of agricultural residues has tremendous negative environmental implications including loss of soil nutrients, soil erosion, and release of climate pollutants including greenhouse gasses [48, 52, 55–60]. Added to these are considerable adverse human and animal health concerns (**Figure 2**) [59–61].

Current agricultural residue and waste management models (the circular and bio-economy models) (**Figures 3** and **4**) emphasize profitability, sustainability, technical feasibility, and adoption potential. That is, an integrated circular and bioeconomy approach that is sustainable, up scalable, crop- and region-specific, socially inclusive, environmentally sound, and technically robust [52]. The approach further harnesses the synergies existing among alternative options, aims at mitigating climate change, and contributes to achievement of sustainable development goals [52].

This perspective is hence driven by the urge for better human and animal health and welfare, sustainable agricultural and animal production, the need to decarbonize the agricultural economy, exploitation of emerging opportunities

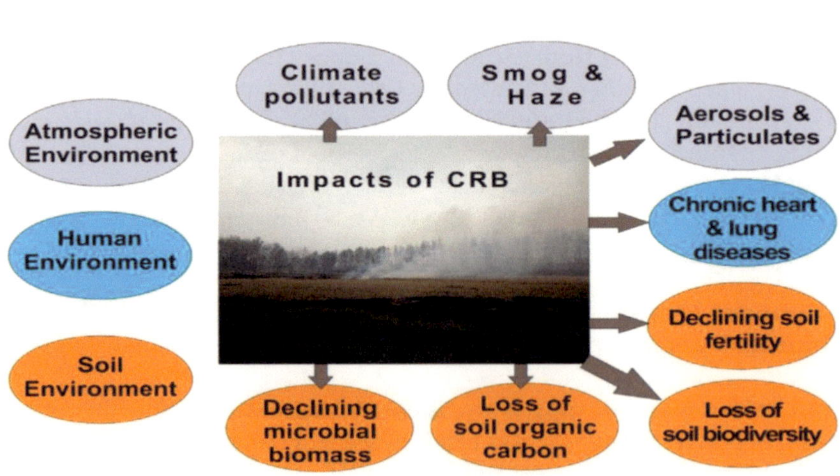

Figure 2.
Impact of crop residue burning (CRB) [52].

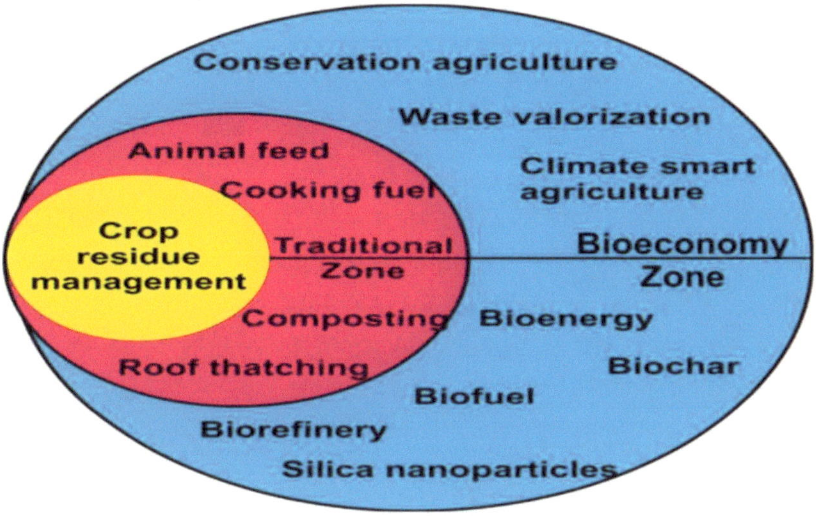

Figure 3.
Crop residue management: A paradigm shift from the traditional approach to agricultural bioeconomy and circular economy [52].

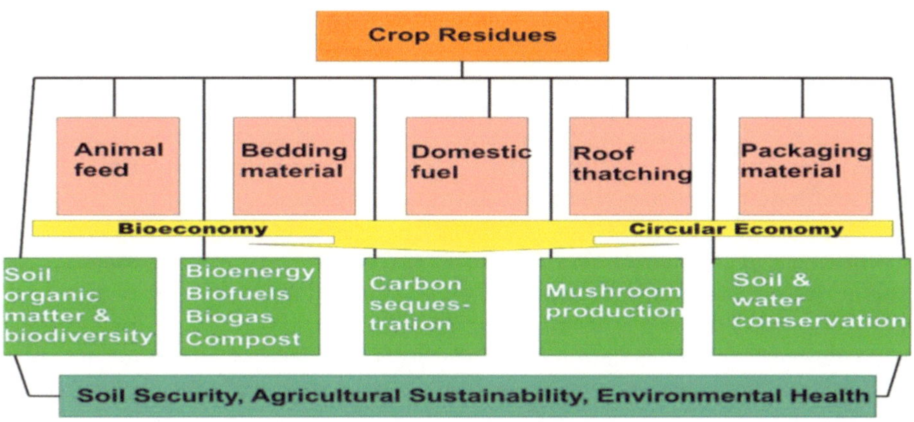

Figure 4.
Crop residue management intervention: Bioeconomy and circular economy perspectives [52].

through waste valorization and production of value-added products, to end waste burning and derive optimal economic benefits from agricultural residues and food wastes for improved livelihood, circularity in agricultural production system for example through reuse and recycling of residues, and the need to mitigate climate change (**Figure 5**) [52, 62].

The circular and bioeconomy approaches have become critical in the backdrop of declining soil fertility, and productivity, need of food and nutritional security, carbon sustainability, and to mitigate adverse health effects and greenhouse gas emission [52, 62]. Key components of the agricultural circular and bioeconomy waste management interventions are conservation agriculture, in situ crop residue incorporation, biomass energy production, biofuel generation, biochar and activated carbon production, residue composting and biofertilizer production, and substrate for edible fungi cultivation [52, 63–66], production of materials such as packagings, food coatings,

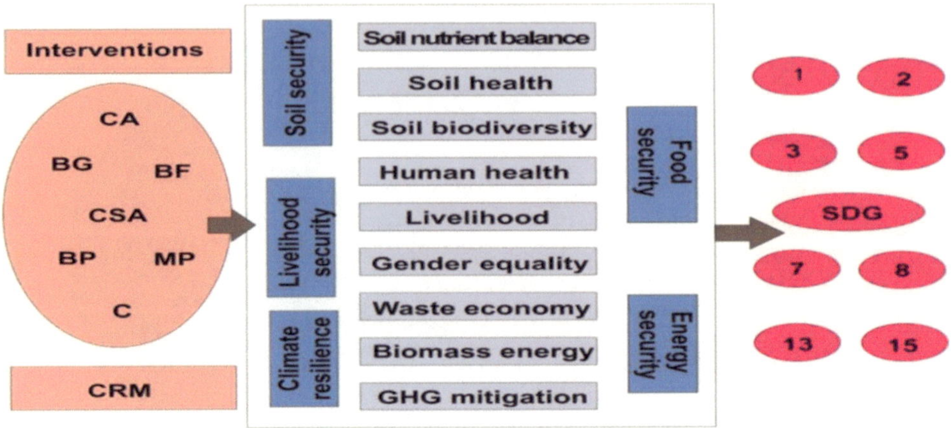

Figure 5.
Mapping of benefits from bioeconomy-driven crop residue management. CA: conservation agriculture; CSA: climate smart agriculture; C: composting; BF: biofuel; BP: biochar production; BG, biogas; MP: mushroom production [52]. Numbers represent targeted sustainable development goals (SDGs).

biopesticides, single cell proteins, and animal proteins [13, 67], and extraction of bioactive molecules and functional groups for varied applications [47, 48, 67]. Obviously, the potential availability and sustainability of feedstock supply for particular industry would depend on alternative or competing uses, cost of production and supply, crop yield and revenue from crops to farmers, and scale of crop farming enterprise [1, 11, 68]. Following is a review of application and allocation of agro-industrial residues (crop residues, fruits and vegetable wastes, animal manure, carcasses, and dairy wastes) and food wastes to major competing applications.

3.1 Application of agro-industrial wastes to soil conservation, soil erosion control, and maintenance of soil fertility

Soil conservation and enrichment are among the traditional and modern applications of crop residues and recyclable agro-industrial wastes including animal manure. It is a vital component of conservation agriculture. Traditionally, some quantities of crop residues are left on-farm to consciously or inadvertently prevent or minimize soil erosion, increase soil organic matter, maintain soil organic carbon, improve soil structure, conserve soil moisture, recycle plant nutrients, maintain soil organisms and microbial population, and maintain soil fertility while some proportions are composted or converted to bio-fertilizer for soil enrichment [7, 33, 69]. The circular and bioeconomy models further recommend in situ crop residue incorporation and no- or minimal-till cultivation practices. In China, it is estimated that 15% of annual residues production are converted to fertilizer while 31% are left on-farm [37, 70]. Wang et al. [71] had reported that crop residue use as fertilizer in Henan Province of China is made up of straw left in field, straw returned to field, straw chipping and mulching in field, quick composting in field, straw pile fermentation returned to field, straw-produced organic fertilizer, straw-produced ammoniation, and straw-produced silage. It is reported that in 2009, 15.4 Mt of wheat straw and 9.7 Mt of rice straw were used for fertilizer in China [72]. Also, crop residues accounted for 12–19% of total organic fertilizer resources [73], provided about 25–35% N, P, and K nutrients, and improved nutrient recycling in the soil. Batidzirai et al. [7] reported that in South Africa, out of a gross amount of 16 Mt yr^{-1} of maize stover, 6.3 Mt are below ground level and serve for soil organic carbon maintenance while 4.2 Mt is required for soil erosion control. Similarly, out of 1.8 Mt of wheat straw, 970,000 tons that are below the ground serve for soil organic carbon maintenance. In a study of the pattern of residue biomass use in cereal-sheep production system of North Africa, Ameur et al. [69] reported that 70.4% of farmers in Tunisia retain below 200 kg of crop residue per hectare (ha^{-1}), 15.1% retain between 200 and 500 kg ha^{-1}, while only 14.5% retain > 500 kg ha^{-1} as mulch. Baudron et al. [74] had reported that only 3% of farmers in Ethiopian Rift Valley retain more than 1 t ha^{-1} of crop residues on the farm. A number of studies [50, 69, 74, 75] have highlighted the challenges associated with crop residue allocation to competing applications including soil conservation in conservation agriculture and in crop-livestock integrated systems. Generally, the quantity of crop residue retained for soil conservation in integrated crop-livestock systems depends on scale of farming (acreage), type of crop, stocking density of livestock, share of livestock income, and crop yield, among other drivers of crop residue management [69].

3.2 Agro-industrial wastes for livestock production (feed and bedding material)

Historically, crop residues were used as animal bedding [11] and as supplemental feed for livestock especially in smallholder mixed agricultural systems. Erenstein [76] had observed that crop residues represent a fundamental resource for crop-livestock integration and intensification, along a broad range of smallholder mixed systems. In tropical and subtropical countries, food wastes serve as food for pet animals and feed for monogastric animals especially pigs. Agricultural biomass is a vital dry season feed resource, providing livestock feed when other resources are scarce. Valbuena et al. [50] stated that in semi-arid and arid areas, crop residue is a vital feed resource for livestock production, and there is increasing demand of agricultural biomass for animal feeding due to low availability of alternative resources [69]. Cooper and Laing [20] stated that the application of crop residue for animal feeding and soil fertility maintenance is very important in maintaining balance and functionality in the rural system. Thus, crop residues play important roles for farmer's livelihood, in various contexts and under different levels of resource availability [50]. In Ethiopia, and most other developing countries of Africa and Asia, crop residues and food wastes are used primarily for animal feeding [77]. In India, crop residues are enriched with urea and molasses as fodder for livestock [37]. In China, 31% of annual crop residue yield was applied to animal feeding [37, 70]. In Henan Province alone, Wang et al. [71] reported the application of 6.9 Mt of wheat straw, 6.0 Mt of corn stalk, and 2.2 Mt of peanut shell and leaves as forage in 2009 with a projected application of 4 Mt of agricultural residue to forage production in 2016. In Denmark, close to ½ of total collected crop residues are used for animal bedding and animal feed and fodder [1]. An estimated 3.6 Mt of dry crop residue was used for animal bedding in Canada in 2011 aside from the amount applied to supplement forage crops for animal feeding [1, 29]. Krausmann et al. [14] reported that of the 4.4 Tg yr^{-1} of crop residue, 2.9 Tg (66%) is appropriated for fodder, animal bedding, and energy. The authors reported that 10% of harvested residues were used to feed livestock in Europe and North America while 83% was used in South and Central Asia. Wirsenius et al. [78] reported global appropriation of 41% of total crop residue to food systems (i.e., livestock feeds) while Weiser et al. [79] reported that 24% of harvested cereal straw was used for livestock husbandry. In the EU, Scarlat et al. [80] reported that 1/5th to 1/3rd of harvested crop residues were applied to livestock production while Ericsson et al. [81] estimated 1/3rd. In SA,

Batidzirai et al. [7] estimated that 260,000 tons yr^{-1} of maize stover out of the harvestable 9.7 Mt yr^{-1} was needed for cattle feed while 70,000 tons out of the harvestable 870,000 tons of wheat straw was applied to livestock bedding. In sub-Saharan Africa with predominantly small holder crop-livestock integration, crop residues are primarily applied to livestock production (fodder, feed, and bedding) such that allocation to other competing applications including soil conservation is problematic [50, 69, 74, 75].

3.3 Application of agro-industrial residues and food wastes in renewable energy production

Among the renewable sources of energy (hydropower, solar, geothermal, wave, wind, biomass, and tidal), only biomass can be converted to three different forms of energy carrier: solid, liquid, and gaseous biofuels [82]. The most common biofuels are bioethanol, biodiesel, and biogas. Bioethanol is generated from starch, sugar, and lignocellulosic-rich crops (phase I) and from food wastes and crop residues feedstocks (phase II); biodiesel is produced from oil rich crops, food wastes, and agricultural residue feedstocks; while biogas is produced from carbohydrate-rich biomass including animal wastes and manure [33]. Biofuels are used in different applications such as generation of electricity, heating, and powering of machines [33]. Generation of energy products from agricultural biomass has in addition environmental advantages: biomass source is storable, inexpensive, energy-efficient, and environmentally friendly. Biomass energy production and biofuel generation are major components of the circular and bioeconomic agro-industrial residue and food waste management option [52, 64, 65]. Utilization of agricultural residues and food wastes for large-scale modern bioenergy production is gaining attention in many countries [83, 84]. Countries such as Denmark, United Kingdom, Spain, Sweden, China, and India have developed large bioenergy facilities [85, 86]. Leading countries for bioethanol production are USA and Brazil [33]. Key residues are maize stover, wheat straw, rice straw and husks, and bagasse [21, 87, 88]. Bagasse is the most commonly used residue, but USA is pioneering the use of maize stover while Denmark is focusing on straw. A study by Bentsen et al. [16] estimated the theoretical global potential of primary agricultural residues from cereals and sugar cane available for biofuel production at approximately 3.7 billion tons of dry matter annually. Earlier studies [10, 14, 18] had estimated 2.7–3.5 billion tons yr^{-1}. A more recent study by Panoutsou et al. [89] projected a technical potential of crop and agro-industrial residue availability for biofuel production in the EU28, Western Bulkans, Turkey, and Ukraine as 400,000 tons of dry biomass yr^{-1} by 2030. In Denmark, Gylling et al. [90] had projected an increase from 1.4 Mt yr^{-1} to approximately 3.0 Mt yr^{-1} by 2020. Studies by Larsen et al. [91] and Thomsen et al. [92] estimated that 50% of straw resources in Denmark are collected for various purposes, 45–50% of which is used for energy generation. Ericsson and Nilsson [81] and Scarlat et al. [80] reported that 20–40% of crop residues produced in Denmark was applied to energy generation. Annual straw consumption for domestic heat and power was put at approximately 1.4 Mt (fresh weight), and a significant proportion is channeled to biorefinery (biofuel, bioethanol), chemical, and material production [16, 80, 81]. It is projected that about 155 Mt of agricultural residue (including 60 Mt of manure) would be applied to bioenergy production in the United States by 2030 [11] without increasing the agriculture share of land resources [93, 94]. Regional estimates reported by UCS [11] include Texas (19.8 Mt), California (9.2 Mt), Alkansas (10.3 Mt), and Iowa (31 Mt). The “Billion ton annual supply study" [21] and its update [22] reported appropriation of 5.6 Mt (1% of total) of corn stover for energy production out of an annual crop residue production of 550 Mt [16]. US DOE [24] reported that 111 million dry tonnes (Mdt) and 94 Mdt of crop residue can be collected at farm-gate prices of \$60 and \$50, respectively; 3/4th of which are corn stover and 1/5th wheat straw. The study projected that by 2030, about 180 Mdt yr^{-1} residue will be available for biofuel production under the base-line scenario while about 320 Mdt yr^{-1} will be available under the high-yield scenario with 85% being corn stover. In addition are 20–26 Mdt of processing and other wastes at \$40–\$60 per dry ton (dt^{-1}). Animal manure was estimated at 30–60 Mdt at farm-gate price of \$50–\$60 dt^{-1}. IARI [37] reported that of the 700 Mt yr^{-1} of crop residues produced in China, 19% (~133 Mt) was used for energy generation while a theoretical estimate of 930.8 Mt of lignocellulosic biomass was reported for China by Ji [30]. The report also indicated that 13.5 Mt yr^{-1} of residues was theoretically available for biofuel production from 19 potential crops. Lignocellulosic biomass availability for bioenergy generation in Iran was reported to be about 11.33 Mt yr^{-1} by Alavijeh and Yaghmaei [31]. For SA, Batidzirai et al. [7] estimated the sustainable biomass energy potential to range from 400 to 550 PJ excluding energy crops, public grasslands, and roadside grasses. Components include maize and wheat residues (6 Mt yr^{-1} or 104 PJ energy equivalent, that is, 5.1 Mt yr^{-1} of maize stover at 94 PJ and 600,000 tons yr^{-1} of wheat straw at 10 PJ), forestry biomass residue (189 PJ: 1° = 41 PJ, 2° = 17 PJ, 3° = 70 PJ; wood chips = 61 PJ), sugar cane plantation and cane processing or bagasse (19–32 PJ), and organic waste from municipal solid wastes (4.5 Mt at 8 PJ). Barahira et al. [33] reported that 13.5 Mt of crop residues are potentially available for biofuel production in SA. In Nigeria and most other Sub-Saharan Africa, renewable energy production from agro-industrial and food wastes is still in its infancy, and large-scale industrial bioenergy production is yet to be accorded priority. Biomass conversion technologies currently applied at elemental level include physical or mechanical conversion (chipping, grinding, milling, and densification) into solid fuels such as briquette and pellets [39, 95–97]; thermochemical conversion (combustion, pyrolysis, gasification, and liquefaction) to produce biochar, bio-oil, and gas such as methane [39, 98, 99]; and biochemical conversion (anaerobic digestion, fermentation, and transesterification) to produce biogas, ethanol, and biodiesel [100–102]. Jakayinfa et al. [39] estimated the theoretical bioenergy potential of agricultural residues and animal wastes for Nigeria at 5.81 EJ (3.64 EJ from agricultural residues and 2.17 EJ from animal wastes) while the technical potential, based on generalized availability factor of 0.3 [103] (range: 0.0–1.0 depending on type of crop residue), was estimated at 1.74 EJ. For Tanzania, total bioenergy potential of crop residues was put at 5714.0 TJ in 2012 with sugarcane and cassava having the highest potential of 2966.4 and 845.0 TJ yr^{-1}, respectively, while a total of 1397.0 TJ yr^{-1} could be generated from animal waste with cattle having the highest potential of 1,139,074,332 MJ yr^{-1} followed by goat (181,036,476 MJ yr^{-1}). The least bioenergy potential was for poultry manure (7,859 MJ yr^{-1}) [6].

3.4 Agro-industrial wastes as substrates for edible fungi (mushroom) cultivation

A rapidly expanding valorization of agro-industrial residues is as substrates for edible fungi (mushroom) cultivation (**Figure 6**). Mushroom cultivation is seen as a major and sustainable component of modern agricultural residue management protocol and circular agricultural systems [105, 106]. Data on global or country-specific allocation of crop residues to mushroom production are scarce in literature. IEA [1] reported that in Canada, about 1.0 Mdt of crop residues were applied to mushroom and horticultural cultivation in 2011. In China, Wang et al. [71] reported that application of agricultural residues to mushroom production has been commercialized. Total crop residue applied to mushroom cultivation in Henan province in 2009 was 2.44 Mt, which accounts for 3% of crop residue generation and 4% of total crop residue utilization in the province. Of these volumes, wheat straw accounted for 0.68 Mt, rice straw, 0.53 Mt [72]. In Luoyang alone, 52 ha and about 200,000 tons of residues were dedicated to mushroom cultivation annually [71, 72]. In India, Raman et al. [107] reported that a total of 39 kinds of agricultural residues from 26 crops provide valuable resources for mushroom cultivation. According to the report, India produces over 620 Mt yr^{-1} of agricultural residue, a substantial quantity of which could be profitably channeled to mushroom production. In eastern Democratic Republic of Congo, Kaziga et al. [108] reported favorable results of an effort to orientate smallholder farmers to valorize wastes from stable crops through mushroom production.

Figure 6.
Major edible mushrooms on agricultural residue substrates [104].

Global mushroom production has increased tremendously over the past decades from about 0.35 Mt in 1965 to about 3.41 Mt in 2007 [107, 109–111]. In 2015, global mushroom market was worth USD 35 billion [107] while Royse et al. [104] reported that global mushroom market was worth USD 63 billion in 2013 made up of medicinal mushrooms: 38%, wild mushrooms: 8%, and edible mushrooms: 54%; with an annual projected increase by USD 34–60 billion [105]. In addition to contributing to food and nutrition security, spent mushroom substrate (SMS) represents a vital resource to produce high-quality compost for growth of other fungi [112], improve animal nutrition and health [105, 106, 113, 114], produce materials [115–117], and extract enzymes for industries [105, 106, 109] and for soil amendment and bioremediation [106, 114, 118, 119]. China is the global leader in mushroom production with 3,918,300 tons yr^{-1} or 64% of global volume and 85% for oyster mushroom production (*Pleurotus spp.*) [109, 110].

3.5 Derivation of bioactive molecules from agro-industrial residues and food wastes

Agricultural residue and food waste biomass stream are widely acclaimed as valuable resources for bioactive compounds (molecules) and functional products. Valorization of agro-industrial residues into food and feed ingredients, dietary supplements, novel bio-components, nutraceuticals, and pharmaceuticals is a potential huge industry and a vital component of the bio-economic and circular model of agricultural residue management. Bioactive molecules from agricultural and food residues include phenols, polyphenols, non-starch polysaccharides (cellulose, hemi-cellulose, and lignin), oligosaccharides, carotenoids, soluble fibers, terpenoids, proteins, tocopherols, and phytosterols [47, 48, 120, 121]. Animal wastes and residues (dairy waste, sea food waste, slaughter wastes) have high levels of proteins, lipids, and minerals [48, 122, 123]. Prado et al. [124] had observed that animal and vegetable wastes are low-cost materials for bioactive compounds using suitable processes. The extraction of bioactive compounds can be integrated into bioenergy facilities, thus enhancing revenue potentials [67, 125]. These bioactive molecules or metabolites can impact human and animal health, for instance, as antioxidant, immunomodulatory, anti-hypertensive, anti-inflammatory, cholesterol reducing, antiaging, anti-cancer, and antidiabetic agents [48, 126–128]; as well as serve as raw materials for the development of functional ingredients useful in food, cosmetics, chemical, and pharmaceutical industries [48, 67, 129–132]. **Figure 7** shows the main classes of bioactive compounds that are derivable from agro-industrial and food waste biomass.

Traditional and novel approaches for extraction of bioactive molecules from organic agricultural and food wastes (**Figure 8**) include (a)solvent extraction (characterized by low processing cost and ease of operation) [120], (b) microbial fermentation (solid-state fermentation) [48, 70, 134–137], (c) supercritical fluid extraction (SFE) [13, 120, 138, 139], (d) subcritical water extraction (SCWE) (characterized by shorter extraction time, lower solvent cost, high quality product, and eco-compatibility) [13, 120, 138, 140, 141], (e) enzyme assisted extraction (EAE) (uses water as solvent, and employs enzymes such as α-amylase, proteases, chitinase, tanase, cellulase, β-glucosidase, xylanase, β-glucanase, and pectinase, which help to degrade cell wall structure and depolymerize plant cell wall polysaccharides, facilitating the release of linked compounds) [13, 48, 120, 142–144], (f) ultra-sound assisted extraction (UAE) [13, 120, 134, 138, 145–147], pulse electric field-assisted extraction [13], and (g) micro-wave assisted extraction (MAE) (characterized by shorter extraction

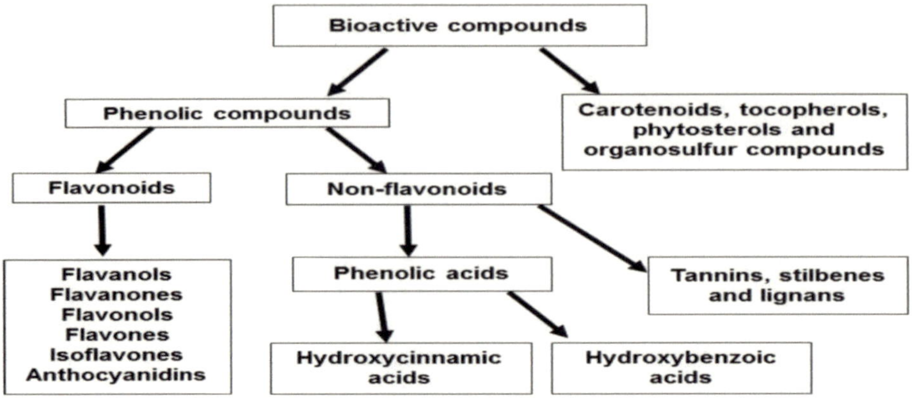

Figure 7.
The main classes of bioactive compounds [133].

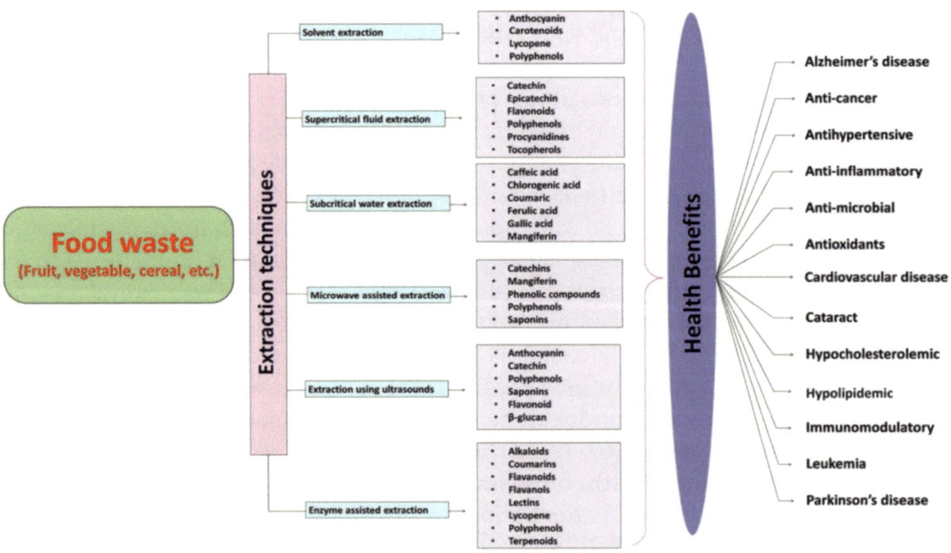

Figure 8.
A schematic representation of different techniques for extraction of bioactive compounds from food wastes and their health benefits [120].

time, higher extraction rate, lesser solvent requirement, and lower cost) [13, 120, 134, 147–149]. Generally, the choice of applied extraction method and recovery rate depends on residue or waste type and bioactive compound of interest [13, 150, 151]. To enhance the efficiency of conventional extraction methods, deep eutectic solvents (DESs) and natural deep eutectic solvents (NADESs) have been suggested to replace organic solvents for extraction of bioactive compounds from agricultural waste biomass materials [13, 152, 153]. Also, the application of encapsulation and nanoemulsion to enhance the stability, bioavailability, and accessibility of derived bioactive compounds is being studied [13, 154, 155].

Khaksar et al. [47] suggested the integration of metabolomics approaches into extraction of bioactive compounds from organic agricultural wastes to gain deeper understanding of the metabolic profile of agro-industrial wastes and enhance their value. Metabolomics approaches include liquid chromatography-mass spectrometry (LC-MS), gas chromatography-mass spectrometry (GC-MS), nuclear magnetic resonance spectroscopy (NMR), and high-performance liquid chromatography (HPLC). Metabolomics had enabled the discovery of over 10,000 different phenolic structures with diverse natures including phenolic acids, flavonoids, and tannins [47, 156].

3.6 Agro-industrial residues as resources for production of materials

Agricultural residues can also be valorized as substrates for production of insects as sources of fat and proteins [67]. Arthropods such as flies and maggots can grow on a variety of organic matter including animal manure, vegetable, and fruit wastes. These arthropods can be converted to animal flour of higher protein content and incorporated as sources of protein and fat in animal feed (livestock, aquaculture, pet) [67]. Larval digestion (bioconversion or biotransformation) of plant and animal materials yield high-value products such as organic fertilizers [67]. In addition, high-value products such as chitin, antibiotics, and peptides with bio-stimulant activity can be extracted for use in animal health and nutrition [67]. Other useful biocompounds derivable from agricultural residues include biopesticides [157, 158], pectin as edible food coating [159, 160], natural aromas [67, 161], quercetin, and flavonols [67, 162]. The replacement of non-recyclable raw materials for production of functional components in industries has become essential as a waste management strategy and in line with circular economic principles. Consequently, valorization of agricultural waste biomass and food wastes to functional materials is being promoted. Growth or culture media, packaging materials, biochar, biopolymers, bioplastics, single cell proteins (from microbial biomass), enzymes, organic acids, biofertilizers, and compost are materials derivable from agricultural residues [13, 67, 134, 163–167]. Biopolymers and bioplastics are highly biodegradable, biofunctional, biostable, and biocompatible and have wide applications in cosmetics, pharmaceutical, chemical, food, and beverage industries [13, 168].

3.7 Valorization of agro-industrial residues through activated charcoal production

Activated biochar or activated carbon or activated charcoal (AC) is a carbonaceous material of high surface area, and large pore volume, widely employed for adsorption of pollutants such as toxins, poisons, metals, and chemicals. Activated carbon production from agricultural residues and food wastes is viewed as a major valorization product with considerable and sustainable potential for application in various fields including bioremediation, water treatment, soil amendment, soil fertility, enhancement of soil biophysical and chemical characteristics, carbon sequestration, medical health, animal health, animal production, horticulture, and climate change mitigation [52, 169, 170]. Biochar production from agro-industrial and food wastes is particularly important in developing countries including Nigeria owing to its simplicity and wide applications. Adding activated carbon to soils could lock away carbon for centuries owing to slow microbial decomposition [171–173]. Minasny et al. [174] submitted that recycling stable crop residues as biochar substantially contribute to carbon sequestration in soils. Biochar addition to soil was shown to increase crop yield by about 25% on average [172, 175] attributed to enhanced cation exchange capacity, soil aggregation, and hydraulic conductivity [176]. In industrial waste water treatment, AC is widely employed for reduction of pollutants such as heavy metals and toxic chemicals (e.g., phenols and their derivatives) [177]. The surface chemistry of activated charcoal confers on it the ability to adsorb many gases, aqueous liquid, chemicals, and poisons [177–179]. Several studies showed that activated charcoal is harmless even when it is accidentally consumed, inhaled, or comes in contact with the skin. When mixed with water and swallowed to counteract poisoning, activated charcoal adsorbs the poison or drug, inactivating it and then carries it inert through the entire length of the digestive tract out of the body [180]. Majewska [181] reported a 3.5–5.0% increase in body weight and higher carcass and organ weights compared with control in broiler chickens on 3% dietary inclusion level of hard wood charcoal. The author attributed the results to the detoxifying effects of charcoal, lowered surface tension of the intestinal digesta, and enhanced liver function with respect to fat digestion. Jiya [182] supplemented activated charcoal at 0.5% in broiler feeds and noted increased relative organ weights and reduced cholesterol level in the carcass attributed to efficient mineral uptake and nutrient utilization. Drunna et al. [183] reported improved growth rate and reduced flatulence, fly population, and litter odor at varied inclusion levels of wood charcoal in feed of broiler chickens. Dim et al. [184] observed improved daily weight gain and feed conversion ratio in broiler chickens fed 6% dietary charcoal inclusion compared with other groups. Linhoss et al. [185] reported positive effects of biochar in litter amendment in broiler production while Schmidt et al. [186] reported that biochar has the potential to improve animal health, feed efficiency, and livestock housing climate; reduce nutrient losses and greenhouse gas emissions; increase soil organic matter content, and thus soil fertility. Zhang et al. [187] observed that biochar remediates organic pollutants by hydrogen binding, surface complexation, electrostatic attractions, and acid-base interactions; and heavy metals in soils by precipitation, surface complexation, chemical reduction, cation exchange, and electrostatic attraction. Biochar can improve cation exchange capacity, neutralize acidic soils, and enhance soil fertility [134, 188, 189]. It enhances organic solid waste decomposition by enhancing microbial population. It removes pollutants such as antibacterial drugs from water and wastewater [190, 191].

Traditionally, AC was produced from coal, lignite, petroleum residue, and hard wood biomass (fossil-related resources) [177, 178]. These materials are costly, exhaustible, and unfriendly to the ecosystem [177, 192]. Today, recyclable agro-industrial residues and food wastes are promoted as viable and sustainable alternatives being renewable, readily available, inexpensive, environmentally friendly, and an additional income to growers, farmers, and vendors [67, 177, 193]. Generally, the use of agricultural wastes such as corn cob, groundnut shell, poultry litter, rice husk, palm kernel shell, and coconut shell in the production of value-added products is gaining momentum [67, 173, 178, 194]. Activated carbon can hence be produced from all plant parts, animal manure, bones, fruit and tuber peels, husks, corn cob, stalk, straw, shell, and fruit stone [52, 67, 178]. These materials are broadly classified as woody materials composed of cellulose, hemicellulose, and lignin, and non-woody biomass composed of cellulose, hemicellulose, lignin, lipids, proteins, sugars, water, hydrocarbons, starch, and many other functional groups such as carbonyl, carboxylic, chromene, ethers, lactone, phenol, pyrone, and quinone groups, which contribute to the physicochemical properties and activity of the final product [178, 195]. Generally, choice of feedstock for AC production is informed by high carbon content, low inorganic matter, high density, and high content of volatile matter, ready availability, and low cost, low rate of degradation during storage, and high AC yield upon pyrolysis [173, 178, 195]. Studies on AC production from bioresources include tomato waste [196], corn cob [197], corn stalk [198], groundnut shell [199], palm kernel shell [200, 201], coconut shell [202, 203], chestnut oak shell [204], peanut shell [147, 205–208], rice husk and straw [209, 210], apple waste [211], grape stalk [212], coffee grounds [213], palm oil mill residue [214, 215], oil palm empty fruit bunch [216, 217], bio-waste mixtures [218], food waste [177], poultry litter [203, 219, 220], pig dung [203, 221], dairy cattle carcass [222], cow dung [203, 223], and chicken feather [224].

The characteristics, properties, and performance of any AC sample would depend on the type of feedstock, temperature, and resident time of pyrolysis (carbonization) [134, 177, 178, 218], the activation process adopted: physical, chemical, or physicochemical [177, 218, 225], as well as the content of inorganic elements and other functional groups [177, 218]. These factors influence the internal pore structure of the AC, which determines its adsorbent capacity [177, 218]. Pores are classified as micropores (< 2nm), mesopores (2–50 nm), and macropores (> 50 nm) [177, 218, 226]. **Figure 9** is a schematic representation of pore structure in activated carbon. The higher the internal pore structure, the more the surface area and adsorbent capacity of the AC [177, 178].

3.7.1 Production process of activated carbon

In principle, AC production involves physical or chemical treatment [178]. Physical treatment includes carbonization of feedstock in the absence of oxygen and in the presence of an inert gas (e.g., argon, Ar) followed by activation of the charcoal (biochar or carbon) using an oxidizing agent (steam, CO_2, or their mixture). Chemical treatment involves carbonization of feedstock to which had been added an activating agent (a strong dehydrating and oxidizing agent such as H_3PO_4, $ZnCl_2$, KOH, NaOH, and H_2SO_4 under nitrogen atmosphere. Thus, two principal steps: carbonization and activation, are involved in AC production [177, 178, 218, 228]. Equipment for AC production varies in sophistication depending on the degree of mechanization.

Carbonization aims to decompose the feedstock, eliminate non-carbon species (volatile matter, non-carbon elements, namely nitrogen, oxygen, hydrogen, sulfur; aromatics, etc.) and deposit biochar (charcoal or carbon) having essentially a fixed carbon content [178] and substantial pore structure. Reported pyrolysis (carbonization) temperature and resident time vary widely probably on account of differences

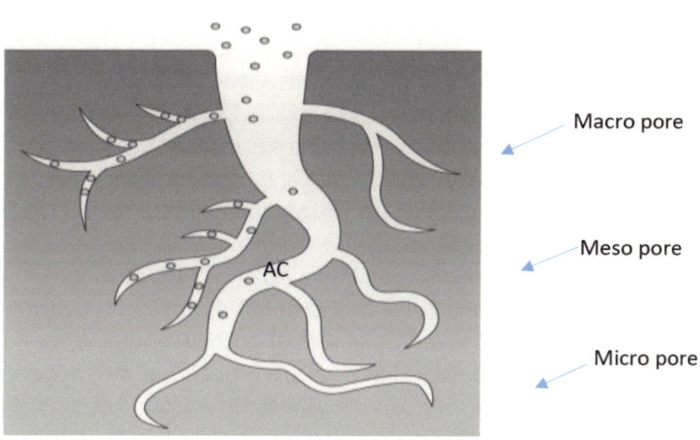

Figure 9.
Pore structures in activated carbon (AC) [227].

in feedstock material, level of sophistication of equipment and automation, as well as other environmental conditions. Generally recommended temperature range is 400–900°C under inert atmosphere [75, 229, 230]. Budi et al. [231], however, reported pyrolysis of coconut shell in a kiln at 75–150°C for 6 h in argon atmosphere. Yasin and Pravinkumar [218] pyrolyzed wood pieces at 400–500°C, coconut shell at 320–400°C, and saw dust at 200–300°C. Yu et al. [177] using a vertical tube furnace pyrolyzed a blend of food wastes at 275–525°C for 30–120 min.

Activation is a key step in AC production, which aims to enhance the critical performance parameters of AC such as pore structure, pore volume, porosity, surface area, fixed carbon, and mineral contents. Activation can be by physical or chemical treatments. Physical activation involves treating biochar with an oxidizing gas or a combination of oxidizing gases such as oxygen, carbondioxide, and steam at high temperatures commonly in the range 500–1000°C [178]. Budi et al. [228] reported physical activation of biochar derived from coconut shell in argon gas furnace at 532, 700, and 868°C for 10–120 min. The activating gas opens previously formed pores blocked by tar or pyroligneous liquids formed during carbonization, creates new pores, and widens existing ones by removal of reactive carbon species, and other volatile contents of the biochar [177, 178]. The reaction(s) during physical activation could involve the below [178]:

a. Oxygen:	b. Carbon dioxide	c. Steam
$C + O_2 = CO_2$	$C + CO_2 = 2CO$	$C + H_2O = H_2 + CO$
$2C + O_2 = 2CO$		$CO + H_2O = H_2 + CO_2$
		$C + 2H_2 = CH_4$

In chemical activation, strong dehydrating and oxidizing chemicals are employed. These include alkali: potassium hydroxide (KOH) [177, 178], potassium carbonate (K_2CO_3) [153, 177, 178, 232], sodium hydroxide (NaOH) [177, 178, 233], and sodium carbonate (Na_2CO_3) [178]; alkali earth metal: aluminum chloride ($AlCl_3$) [178], and zinc chloride ($ZnCl_2$) [177, 178, 230, 231, 233–239]; and acid: phosphoric acid (H_3PO_4) [177, 178, 218], and sulfuric acid (H_2SO_4) [177, 178, 236]. The raw material for biochar production is usually impregnated with the appropriate activating agent prior to carbonization (pyrolysis). The activating agent acts as a dehydrating and/or oxidizing agent, promotes decomposition of the feedstock, inhibits deposition of tar and volatile contents [218] such as pyroligneous liquids, and enhances activated charcoal yield, carbon content, porosity, and surface area. At the end of the process, the resulting activated carbon is washed in acid or alkali depending on the activating agent used in order to remove remaining activating agent lodged in the pore structure [218]. An alternative procedure is to immerse biochar resulting from pyrolysis in the choice activating agent for a time duration of 24 h [231] followed by washing and drying. Esmar Budi et al, [231] reported the application of double activation, which involved chemical activation followed by physical activation using a horizontal furnace at 400°C in argon (Ar) gas (200 kg m^{-3}) environment for I h. Whereas activated charcoal produced by physical activation is neater and requires no further washing that from chemical activation has higher surface area and pore volume. Chemical activation also requires lower temperature commonly of range 450–600°C [177]. Among the chemical activators, $ZnCl_2$ and H_3PO_4 are the most applied in the industry [177]. The atmosphere for chemical activation is either inert gas (e.g., Ar) or air [177].

A number of studies have evaluated the physicochemical properties of AC derived from a wide range of recyclable biological waste materials under varied pyrolysis conditions, activation agents, and time duration. Budi et al. [228] studied the effect of activation temperature (532, 700, and 868°C), and resident activation time (10, 60, and 120 min) under argon gas pressure (6.59, 15, and 23.4 kg/cm^2) on pore structure and carbon content of coconut-shell-derived biochar and reported decreases in pore size but increases in pore volume and uniform pore distribution with increasing activation temperature, resident activation time, and gas pressure. Resident activation time alone did not influence pore size and pore distribution. The authors attributed the increase in pore volume to formation of new pores especially micro pores due to release of more volatile components from the biochar, as well as decreased pore size. Increase in pore volume increases the surface area of the activated carbon sample. The authors also reported increase in carbon content, which was attributed to the increase in pore numbers with increase in activation temperature, which caused the release of more volatile components. Esmar Budi et al, [231] evaluated the effects of chemical and physical activation on coconut-shell-derived charcoal. Chemical activation was performed with KOH (30, 40, 50, and 60%), NaOH (1, 2, 4, 7, and 11%), HCl (2, 4, and 6%), and H_3PO_4 (2, 4, and 6%) for 24 h while physical activation was with steam at 400°C in argon gas environment (200 kg m^{-3}) for 1 h. There were increases in pore size; pore number initially increased but decreased as chemical concentration increased. Carbon content decreased with increasing chemical concentration. The decrease in pore number was linked to excessive widening of pre-formed pores, coalesce of pores, and collapse of carbon structure due to excessive chemical attack. The authors also noted an increase in total surface area, which was attributed to formation of new micro pores, and widening of existing pores. Yasin and Pravinkumar [218] investigated the effects of pyrolysis parameters and chemical (phosphoric acid, H_3PO_4) activation on properties of charcoal derived from some bio-waste materials. Biochar yield was higher with increasing pyrolysis temperature and resident time, and yield varied with type of starting material with coconut shell (56.66–63.33%) > saw dust (50.00–56.66%) > wood piece (46.66–53.33%%) at all temperatures and time duration. The result was attributed to higher loss of volatile components in the feedstock and higher oxidization of reactive carbon species at the optimal pyrolysis temperature and time duration. Activated charcoal adsorbent capacity was positively correlated with activation temperature and biochar from coconut shell had higher values than that from other materials (1.95–2.67 vs 1.73–1.97 vs 1.45–1.92 mg g^{-1} for coconut shell vs saw dust vs wood piece). Bulk density did not vary clearly with activation temperature, but values were highest for wood piece (0.75–0.78 gm/cm^3) compared with coconut shell (0.56–0.58 gm/cm^3) and saw dust (0.18–0.21 gm/cm^3). Moisture content followed a similar trend as bulk density being highest in wood piece (4.35–5.25%) compared with coconut shell (2.21–3.10%) and saw dust (1.07–1.25%). Volatile matter was highest in saw-dust-activated charcoal (12.60–18.10%) compared with wood piece and coconut shell (11.25–12.35 and 7.30–7.40%, respectively) while ash content was highest in biochar from wood piece (6.95–7.85%) compared with saw dust (5.95–6.15%) and coconut shell biochar (1.50–1.75%). Biochar from coconut shell had the highest content of fixed carbon (88.10–88.88%) compared with saw dust (74.53–80.18%) and wood piece (74.55–77.45%).High bulk density, moisture, and volatile matter are undesirable properties of activated charcoal. A good activated carbon should be highly porous (high total pore volume), of high fixed carbon content and surface area, which enables high adsorption capacity [177]. In a study that evaluated the physicochemical properties and adsorption capacity of biochar and physically (steam) activated biochar derived from a complex of edible food waste feedstock, Yu et al. [177] observed increased biochar yield with increasing pyrolysis temperature and longer pyrolysis resident time. Carbon content also increased with pyrolysis temperature, but this became stable after 60 min pyrolysis duration. Activated carbon yield reduced with higher activation temperature and longer activation time. Before activation, biochar surface area was 10 m^2 g^{-1}, micro pore volume was 0.004 cm^3 g^{-1}, and total pore volume was 0.016 cm^3 g^{-1}. These indices ranged between 288 and 745 m^2 g^{-1}, 0.072 and 0.196 cm^3 g^{-1}, and 0.160 and 0.792 cm^3 g^{-1}, respectively, after activation. The activation temperature and time duration for the highest activated carbon surface area (745 m^2 g^{-1}), micro pore volume (0.196 cm^3 g^{-1}), and total pore volume (0.792 cm^3 g^{-1}) were 950°C and 1 h, 850°C and 3h, and 950°C and 5 h, respectively. Thus, surface area and total pore volume increased with activation temperature while micro pore volume reached maximum value at 850°C. Biochar and activated carbon produced at 750°C had very small volume of meso and macro pores but higher activation temperatures (850 and 950°C) produced enlarged pores yielding more meso and macro pores. Longer activation time reduced surface area and micro pore volume but increased total pore volume at 950°C. Larger pores were obtained with increased activation time, thereby reducing number of micro pores. Because larger pores have relatively smaller surface area, converting micro pores to meso and macro pores will reduce activated carbon surface area. The authors reported that carbon content was 48.8% in the feedstock, 71.9% in derived biochar (pyrolysis temperature, 525°C for 2 h), and 68.8% after activation at 750°C for 3 h, 62.6% at 850°C for 3 h, 40.9% at 950°C for 3 h, 48.0% at 950°C for 1 h, and 33.2% at 950 for 5 h clearly showing reduced fixed carbon with increasing activation temperature and activation duration. Nitrogen (N), H, and S content followed a similar trend as fixed carbon. Sodium (Na), Ca, and P increased with activation temperature and time duration and were generally higher in activated sample than in biochar and feedstock. It does appear, hence, that increased activation temperature and time duration preserved the mineral content of biochar.

4. Conclusion and future perspective

With rising global human and animal population, demand for food, and food production will continue to increase leading to increases in waste generation and negative environmental challenges. Sustainable agricultural production and agro-industrial processing, environmental, human, animal, and climate health depend substantially on effective waste management. Circular agricultural production and bioeconomic agro-industrial waste management models are key to achieving the vision to considerably reduce waste generation, reuse wastes, and recycle wastes. Continued intellectual brainstorming and research will enable the arrival at the goal of turning wastes to wealth and an agricultural production system that produces nil wastes.

DOI: http://dx.doi.org/10.5772/intechopen.109181

Author details

Cosmas Chikezie Ogbu* and Stephen Nnaemeka Okey
Department of Veterinary Biochemistry and Animal Production, College of Veterinary Medicine, Michael Okpara University of Agriculture Umudike, Abia State, Nigeria

*Address all correspondence to: ogbu.cosmas@mouau.edu.ng

References

[1] IEA (International Energy Agency). Mobilization of agricultural residues for bioenergy and higher value bio-products: Resources, barriers and sustainability. IEA Bioenergy Task. 2017;**43**:1

[2] Torén J, Wirsenius S, Anttila P, Böttcher H, Dees M, Ermert J, et al. Biomass Energy Europe: Executive Summary, Evaluation and Recommendations. Freiburg: University of Freiburg; 2011

[3] Sadh PK, Duhan S, Duhan JS. Agro-industrial waste and their utilization using solid state fermentation: A review. Bioresources and Bioprocessing. 2018;**5**:1

[4] Mendu V, Shearin T, Campbell JE Jr, Stork J, Jae J, Crocker M, et al. Global bioenergy potential from high-lignin agriculture residue. PNAS. 2012;**109**(10):4014-4019. DOI: 10.1073/pnas.1112757109

[5] Biomass Energy Europe. Status of biomass resource assessmentsversion 3. 2010

[6] Lyakurwa FS. Assessment of the energy potential of crop residues and animal waste in Tanzania. Independent Journal of Management and Production. 2016;**7**(4):1227-1239

[7] Batidzirai B, Valk M, Wicke B, Junginger M, Oaioglu V, Euler W, et al. Current and future technical, economic and environmental feasibility of maize and wheat residue supply for biomass energy application: Illustrated for South Africa. Biomass and Bioenergy. 2016;**92**:106-129

[8] Kemausuor F, Kamp A, Thomsen ST, Bensah EC, Østergård H. Assessment of biomass residue availability and bioenergy yields in Ghana. Resources, Conservation and Recycling. 2014;**86**:26-37

[9] Belewu MA, Babalola FT. Nutrient enrichment of some waste agricultural residues after solid state fermentation using *Rhizopusoligosporus*. Journal of Applied Bioscience. 2009;**13**:695-699

[10] Lal R. World crop residues production and implications of its use as a biofuel. Environment International. 2005;**31**(4):575-584

[11] UCS (Union of Concerned Scientists). Turning Agricultural Residues and Manure into Bioenergy. Cambridge: UCS; 2014

[12] Cho EJ, Trinh LTP, Song Y, et al. Bioconversion of biomass waste into high value chemicals. Bioresource Technology. 2020;**298**:122386. DOI: 10.1016/j.biortech.2019.122386

[13] Capanoglu E, Nemli E, Tomas-Barberan F. Novel approaches in the valorization of agricultural wastes and their applications. Journal of Agricultural and Food Chemistry. 2022;**70**(23):6787-6804. DOI: 10.1021/acs.jatc.1co7104

[14] Krausmann F, Erb K-H, Gingrich S, Lauk C, Haberl H. Global patterns of socioeconomic biomass flows in the year 2000: A comprehensive assessment of supply, consumption and constraints. Ecological Economics. 2008;**65**:471

[15] Rogner H-H, Aguilera RF, Bertani R, Bhattacharya SC, Dusseault MB, Gagnon L, et al. Chapter 7 - Energy Resources and Potentials. Global Energy Assessment - Toward a Sustainable Future. Cambridge, UK: Cambridge University Press; 2012. pp. 423-512

[16] Bentsen NS, Felby C, Thorsen BJ. Agricultural residue production and potentials for energy and material

services. Progress in Energy and Combustion Science. 2014;**40**:59-73

[17] Smil V. Crop residues: Agriculture's largest harvest - crop residues incorporatemore than half of the world agricultural phytomass. Bioscience. 1999;**49**:299

[18] Hakala K, Kontturi M, Pahkala K. Field biomass as global energy source. Agricultural and Food Science. 2009;**18**:347

[19] Food and Agricultural Organization. FAOStats 2001. Rome: FAO; 2001

[20] Cooper CJ, Laing CA. A macro analysis of crop residue and animal waste as a potential energy source in Africa. Journal of Energy in Southern Africa. 2007;**18**(1):10-19. DOI: 10.17159/2413-3051/2007/v 18i/93339

[21] Perlack RD, Wright LL, Turnhollow AF, Graham RL, Stokes BJ, Erbach DC. Biomass as Feedstock for a Bioenergy and Bioproducts Industry: The Technical Feasibility of a Billion-ton Annual Supply. Oak Ridge TN: Oak Ridge National Laboratory; 2005

[22] U.S. Department of Energy. U.S. billion-ton Update: Biomass Supply for a Bioenergy and Bioproducts Industry. Oak Ridge National Laboratory: Oak Ridge, TN, USA; 2011. p. 227

[23] Chatterjee A. Annual crop residue production and nutrient replacement costs for bioenergy feedstock production in United States. Agronomy Journal. 2013;**105**:685

[24] U.S. DOE. 2016 Billion-Ton Report: Advancing Domestic Resources for a Thriving Bioeconomy. Oak Ridge, TN, USA: Oak Ridge National Laboratory; 2016

[25] Kim S, Dale BE. Global potential bioethanol production from wasted crops and crop residues. Biomass and Bioenergy. 2004;**26**(4):361-375

[26] Bloomberg. Biomass Supply Costs – Fields of Gold? London: UK, Bloomberg New Energy Finance; 2011

[27] de Wit M, Faaij A. European biomass resource potential and costs. Biomass and Bioenergy. 2010;**34**(2):188-202

[28] Energinet.dk. Energikoncept 2030 - Baggrundsrapport [Energy concept 2030 - Background report]. Fredericia, DK, Energinet.dk. 2015

[29] Li X, Mupondwa E, Panigrahi S, Tabil L, Sokhansanj S, Stumborg M. A review of agricultural crop residue supply in Canada for cellulosic ethanol production. Renewable and Sustainable Energy Reviews. 2012;**16**(5):2954-2965

[30] Ji L-Q. An assessment of agricultural residue resources for liquid biofuel production in China. Renewable and Sustainable Energy Reviews. 2015;**44**:561-575

[31] Alavijeh MK, Yaghmaei S. Biochemical production of bioenergy from agricultural crops and residue in Iran. Waste Management. 2016;**52**:375-394

[32] Pradhan A, Mbohwa C. Development of biofuels in South Africa: Challenges and opportunities. Renewable and Sustainable Energy Reviews. 2014;**39**:1089-1100

[33] Barahira DS, Okudoh VI, Eloka-Eboka AC. Suitability of crop residues as feedstock for biofuel production in South Africa: A sustainablewin-win scenario. Journal of Oleo Science. 2021;**2021**:1-14. DOI: 10.5650/jos.ess20288

[34] Devi S, Gupta C, Jat SL, Parmar MS. Crop residue recycling for economic and

environmental sustainability: The case of India. Open Agriculture. 2017;**2**:486-494. DOI: 10.1515/opag-2017-0053

[35] GoI. Annual Report 2016. New Delhi: Ministry of New and Renewable Energy; 2016

[36] Pathak BS. Crop Residue to Energy. In: Chadha KL, Swaminathan MS, editors. Environment and Agriculture. New Delhi: Malhotra Publishing House; 2004. pp. 854-869

[37] IARI. Crop Residue Management with Conservation Agriculture: Potential, Constraints and Policy Needs. New Delhi: Indian Agricultural Research Institute; 2012. p. 32

[38] FAO. Crop Residue and Agro-industrial By-products in West Africa: Situation and Way Forward for Livestock Production. Rome: Food and Agriculture Organization; 2014

[39] Jekayinfa SO, Orisaleye JI, Pecenka R. An assessment of potential resources for biomass energy in Nigeria. Resources. 2020;**9**:92-132. DOI: 10.3390/resources9080092

[40] Ben-Iwo J, Manovic V, Longhurst P. Biomass resources and biofuels potential for the production of transportation fuels in Nigeria. Renewable and Sustainable Energy Reviews. 2016;**63**:172-192. DOI: 10.1016/j.rser.2016.05.050

[41] Agba MM, Ushie EF, Abam I, Agba MS, Okoro J. Developing the biofuel industry for effective rural transportation. European Journal of Scientific Research. 2010;**40**:441-449

[42] Simonyan K, Fasina O. Biomass resources and bioenergy potentials in Nigeria. African Journal of Agriculture. 2013;**8**:4975-4989. DOI: 10.5897/AJAR2013.6726

[43] Ishola MM, Brandberg T, Sanni SA, Taherzadeh MJ. Biofuels in Nigeria: a critical and strategic evaluation. Renewable Energy. 2013;**55**:554-560. DOI: doi.org/10.1016/j.renene.2012.12.021

[44] Okeh OC, Onwosi CO, Odibo FJC. Biogas production from rice husks generated from various rice mills in Ebonyi State, Nigeria. Renewable Energy. 2014;**62**:2048

[45] Akorode MF, Ibrahim O, Amuda SA, Otuoze AO, Olufeagba BJ. Current status and outlook of renewable energy development in Nigeria. Nigerian Journal of Technology. 2017;**36**(6):196-212

[46] Harris DR, Hillman GC. Foraging and Farming: The Evolution of Plant Exploitation. London, UK: Routledge; 2014

[47] Khaksar G, Sirijan M, Suntichaikamolkul N, Sirikantaramas S. Metabolomics for agricultural waste valorization: Shifting toward a sustainable bioeconomy. Frontiers in Plant Science. 2022;**13**:938480. DOI: 10.3389/fpls.2022.938480

[48] Lemes AC, Egea MB, de Oliveira Filho JG, Gauterio GV, Ribeiro BD, Coelho MAZ. Biological approaches for extraction of bioactive compounds from agro-industrial byproducts: A review. Frontiers in Bioengineering and Biotechnology. 2022;**9**:802543. DOI: 10.3389/fbioe

[49] Dini I, Izzo L, Graziani G, Ritieni A. The nutraceutical properties of Pizza napoletana marinara TSG a traditional food rich in bioaccessible antioxidants. Antioxidants. 2021;**10**:495

[50] Valbuena D, Tui SH-K, Erenstein O, Teufel N, Duncan A, Abdoulaye T, et al. Identifying determinants, pressures and trade-offs of crop residue use in mixed smallholder farms in Sub-Saharan Africa

and South Asia. Agricultural Systems. 2015;**134**:107-118

[51] Lohan S, Jat H, Yadav A, Sidhu H, Jat M, Choudhary M, et al. Burning issues of paddy residue management in north-west states of India. Renewable and Sustainable Energy Reviews. 2018;**81**:693-706

[52] Venkatramanan V, Shah S, Rai AK, Prasad R. Nexus between crop residue burning, bioeconomy and sustainable development goals over North-Western India. Frontier in Energy Research. 2021;**8**:614212. DOI: 10.3389/fenrg.2020.614212

[53] Lopes AA, Viriyavipart A, Tasneem D. The role of social influence in crop residue management: Evidence from Northern India. Ecological Economics. 2020;**169**:106563

[54] Ravindra K, Singh T, Mor S. Emissions of air pollutants from primary crop residue burning in India and their mitigation strategies for cleaner emissions. Journal of Cleaner Production. 2019;**208**:261-273. DOI: 10.1016/j.jclepro.2018.10.031

[55] Chawala P, Sandhu HAS. Stubble burn area estimation and its impact on ambient air quality of Patiala & Ludhiana district, Punjab, India. Heliyon. 2020;**6**:e03095. DOI: 10.1016/j.heliyon.2019.e03095

[56] Jat HS, Jat RD, Nanwal RK, Lohan SK, Yadav AK, Poonia T, et al. Energy use efficiency of crop residue management for sustainable energy and agriculture conservation in NW India. Renewable Energy. 2020;**155**:1372-1382. DOI: 10.1016/j.renene.2020.04.046

[57] Jethva H, Torres O, Field R, Lyapustin A, Gautam R, Kayetha V. Connecting crop productivity, residue fires, and air quality over northern India. Scientific Reports. 2019;**9**:16594. DOI: 10.1038/s41598-019-52799-x

[58] Li R, Chen W, Zhao H, Wu X, Zhang M, Tong DQ, et al. Inventory of atmospheric pollutant emissions from burning of crop residues in China based on satellite-retrieved farmland data. China Geographical Science. 2020;**30**:266-278

[59] Sarkar S, Singh RP, Chauhan A. Increasing health threat to greater parts of India due to crop residue burning. The Lancet Planetary Health. 2018;**2**:e327-e328. DOI: 10.1016/S2542-5196(18)30166-9

[60] Sokhansanj S, Yadav YK, Lau A, Yadvika ME, Verma K. Identification of techno-economic viable crop residue utilization in India. Journal of Sustainable Bioenergy Systems. 2021;**11**:94-104. DOI: 10.4236/jsbs.2021.112007

[61] Chakrabarti S, Khan M, Kishore A, Roy D, Scott S. Risk of acute respiratory infection from crop burning in India: Estimating disease burden and economic welfare from satellite and national health survey data for 250 000 persons. International Journal of Epidemiology. 2019;**48**:1113-1124. DOI: 10.1093/ije/dyz022

[62] Sarkar S, Skalicky M, Hossain A, Brestic M, Saha S, Garai S, et al. Management of crop residue for improving input use efficiency and agricultural sustainability. Sustainablility. 2020;**12**:9808. DOI: 10.3390/su/2239808

[63] Chivenge P, Rubianes F, Van Chin D, Van Thach T, Khang VT, Romasanta RR, Van Hung N et al. Rice straw incorporation influences nutrient cycling and soil organic matter. Sustainable Rice Straw Management Gummert, N. Van Hung, P. Chivenge, and B.

Douthwaite (ed.) Switzerland, Cham: Springer; 2019. pp. 131-144. DOI: 10.1007/978-3-030-32373-8_8

[64] Ngan N, Chan F, Nam T, Van Thao H, Maguyon-Detras MC, Hung DV, Cuong DM. Anaerobic digestion of rice straw for biogas production. In: Sustainable Rice Straw Management. Gummert, M., N. Van Hung, P. Chivenge, and B. Douthwaite (ed.) Switzerland, Cham: Springer; 2019. pp. 65-92

[65] Prasad S, Kumar S, Sheetal K, Venkatramanan V. Global climate change and biofuels policy: Indian perspectives. In: Venkatramanan V, Shachi S, editors. Global Climate Change and Environmental Policy: Agriculture Perspectives. Singapore, Asia: Springer Nature Singapore Pte Ltd; 2020. pp. 207-226

[66] Shyamsundar P, Springer NP, Tallis H, Polasky S, Jat ML, Sidhu HS, et al. Fields on fire: Alternatives to crop residue burning in India. Science. 2019;**365**:536-538. DOI: 10.1126/science.aaw4085

[67] Grade S, Berling E, Blanco B, Caceres R, Checa JG, Di Lonardo S, et al. Circularity and/or valorization of biomass: crop residue, by-products and extraction of materials. Circular Horticulture. 2019;**2019**:1-14

[68] Muth DJ, Bryden KM. An integrated model for assessment of sustainable agricultural residue removal limits for bioenergy systems. Environmental Modelling & Software. 2013;**39**:50-69

[69] Ameur W, Frija A, Abdeladhim MA, Thabet C. Patterns of use of residue biomass in cereal-sheep production systems of North Africa: Case of Tunisia. Agriculture. 2021;**11**:612. DOI: 10.3390/agriculture 11070612

[70] Jiang D, Zhuang D, Fu J, Huang Y, Wen K. Bioenergy potential from crop residues in China: Availability and distribution. Renewable and Sustainable Energy Reviews. 2012;**16**:1377-1382

[71] Wang Z-W, Lei T-Z, Yan X-Y, Li Y-L, He X-F, Zhu J-L. Assessment and utilization of agricultural residues in HenanProvince China. BioResources. 2012;7(3):3847-3861

[72] HPDRC. Comprehensive Utilization of Agricultural Residues in Henan Province. Zhengzhou, China: Henan Province Development Commission; 2011

[73] Huang HX, Li ST, Li XL. Analysis on the status of organic fertilizer and its development strategies in China. Soil Fertilizer (TurangFeiliao). 2006;**1**:3-8

[74] Baudron F, Delmotte S, Corbeels M, Herrera JM, Tittonell P. Multi-scale trade-off analysis of cereal residue use for livestock feeding vs. soil mulching in the Mid-Zambezi Valley, Zimbabwe. Agricultural System. 2015;**134**:97-106

[75] Jaleta M, Kassie M, Shiferaw B. Tradeoffs in crop residue utilization in mixed crop–livestock systems and implications for conservation agriculture. Agricultural Systems. 2013;**121**:96-105

[76] Erenstein O. Crop residue mulching in tropical and semi-tropical countries: An evaluation of residue availability and other technological implications. Soil and Tillage Research. 2002;**67**:115-133

[77] Alemu T, Chairatanayuth P, Vijchulata P, Tudsri S. Production and utilization of crop residues in three agroecological zones of Eastern Shoa zone, Ethiopia. Kasetsart Journal (Natural Science). 2006;**40**:643-651

[78] Wirsenius S. The biomass metabolism of the food system: A model-based survey of the global and regional turnover of food biomass. Journal of Industrial Ecology. 2003;7:47

[79] Weiser C, Zeller V, Reinicke F, Wagner B, Majer S, Vetter A, et al. Integrated assessment of sustainable cereal straw potential and different straw-based energy applications in Germany. Applied Energy. 2013;**114**:749-762. DOI: 10.1016/j.apenergy.2013.07.016

[80] Scarlat N, Martinov M, Dallemand J-F. Assessment of the availability of agricultural crop residues in the European Union: Potential and limitations for bioenergy use. Waste Management. 2010;**30**:1889

[81] Ericsson K, Nilsson L. Assessment of the potential biomass supply in Europe using a resource-focused approach. Biomass and Bioenergy. 2006;**30**:1

[82] Patel M, Zhang X, Kumar A. Techno-economic and life cycle assessment on lignocellulosic biomass thermochemical conversion technologies: A review. Renewable Sustainable Energy Review. 2016;**53**:1486-1499

[83] Kretschmer B, Allen B, Hart K. Mobilising Cereal Straw in the EU to Feed Advanced Biofuel Production. London: IEEP; 2012

[84] Urosevic DM, Gvozdenac-Urosevic BD. Comprehensive analysis of a strawfiredpower plant in the province of Vojvodina. Thermal Science. 2012;**16**:S97eS106

[85] Peidong Z, Yanli Y, Yongsheng T, Xutong Y, Yongkai Z, Yonghong Z, et al. Bioenergy industries development in China: Dilemma and solution. Renewable and Sustainable Energy Reviews. 2009;**13**:2571e2579

[86] Purohit P. Economic potential of biomass gasification projects under clean development mechanism in India. Journal of Cleaner Production. 2009;**17**:181e193

[87] Chum H, Faaij A, Moreira J, Berndes G et al. Bioenergy. In: IPCC Special Report on Renewable Energy Sources and Climate Change Mitigation [O. Edenhofer, R. Pichs-Madruga, Y. Sokona, K. Seyboth, P. Matschoss, S. Kadner, T. Zwickel, P. Eickemeier, G. Hansen, S. Schlömerand, C. von Stechow (eds)]. Cambridge University Press: Cambridge; 2011

[88] Kline KL, Oladosu GA, Wolfe AK, Perlack RD, Dale VH, McMahon M. Biofuel Feedstock Assessment for Selected Countries. Oak Ridge TN: Oak Ridge National Laboratory; 2008. p. 210

[89] Panoutsou C, Langeveld H, Vis M, Lammens TM, Askew M, Carrez D et al. D8.2 Vision for 1 billion dry tonnes D8.2 Vision for 1 billion dry tonnes lignocellulosic biomas as a contribution to biobased economy by 2030 in Europe. S2Biom. 2016. Available from: https://edepot.wur.nl/517794

[90] Gylling M, Jørgensen U, Bentsen NS, Kristensen IT, Dalgaard T, Felby C, et al. The + 10 Million Tonnes Study: Increasing the Sustainable Production of Biomass for Biorefineries. Copenhagen: Department of Food and Resource Economics, University of Copenhagen; 2013

[91] Larsen SU, Bruun S, Lindedam J. Straw yield and saccharification potential for ethanol in cereal species and wheat cultivars. Biomass and Bioenergy. 2012;**45**:239

[92] Thomsen TP, Ravenni G, Holm JK, Ahrenfeldt J, Hauggaard-Nielsen H, Henriksen UB. Screening of various low-grade biomass materials for low temperature gasification: Method development and application. Biomass and Bioenergy. 2015;**79**:128-144

[93] UCS (Union of Concerned Scientists). The Promise of Biomass:

Clean Power and Fuel—If Handled Right. Cambridge: UCS; 2012

[94] USDA (U.S. Department of Agriculture). Census of Agriculture: Summary and State Data. Washington, DC: USDA; 2009

[95] Jekayinfa SO, Pecenka R, Orisaleye JI. Empirical model for prediction of density and water resistance of corn cob briquettes. International Journal of Renewable Energy Technology. 2019;**10**:212-228

[96] Orisaleye JI, Jekayinfa SO, Adebayo AO, Ahmed NA, Pecenka R. Effect of densification variables on density of corn cob briquettes produced using a uniaxial compaction biomass briquetting press. Energy Sources, Part A: Recovery, Utilization, and Environmental Effects. 2018;**40**:3019-3028

[97] Orisaleye JI, Jekayinfa SO, Pecenka R, Onifade TB. Effect of densification variables on water resistance of corn cob briquettes. Agronomy Research. 2019;**17**:1722-1734

[98] Fogarassy C, Toth L, Czikkely M, Finger DC. Improving the efficiency of pyrolysis and increasing the quality of gas production through optimization of prototype systems. Resources. 2019;**8**:182

[99] Tursi AA. Review on biomass: Importance, chemistry, classification, and conversion. Biofuel Research Journal. 2019;**22**:962-979

[100] Adebayo AO, Jekayinfa SO, Ahmed NA. Kinetic study of thermophilic anaerobic digestion of cattle manure in a continuously stirred tank reactor under varying organic loading rate. ARPN Journal of Engineering Applied Science. 2018;**13**:3111-3118

[101] Aigbodion AI, Bakare IO, Fagbemi EA, Abolagba EO, Omonigho B, Ayeke PO, et al. Viability of biogas production from manure/biomass in Nigeria using fixed dome digester. University Journal of Agricultural Research. 2018;**6**:1-8

[102] Dahunsi SO, Oranusi S, Efeovbokhan VE, Zahedi S, Ojediran JO, Olayanju A, et al. Biochemical conversion of fruit rind of Telfairiaoccidentalis (fluted pumpkin) and poultry manure. Energy Sources, Part A: Recovery, Utilization, and Environmental Effects. 2018;**40**:2799-2811

[103] Deng YY, Koper M, Haigh M, Dornburg V. Country-level assessment of long-term global bioenergy potential. Biomass and Bioenergy. 2015;**2015**(74):253-267

[104] Royse DJ, Baars J, Tan Q. Current overview of mushroom production in the world. In: Zied DC, Pardo-Giminez A, editors. Edible and Medicinal Mushrooms: Technology and Applications. Hoboken: JohnWiley& Sons Ltd; 2017. pp. 5-13

[105] Grimm D, Wosten HAB. Mushroom cultivation in the circular economy. Applied Microbiology and Biotechnology. 2018;**102**:7795-7803. DOI: 10.1007/300253-018-9226-8

[106] Hu Y, Mortimer PE, Hyde KD, Kakumyan P, Thongklang N. Mushroom cultivation for soil amendment and bioremediation. Circular. Agricultural Systems. 2021;**1**:11

[107] Raman J, Lee S-K, Im J-H, Oh M-J, Oh Y-L, Jang K-Y. Current prospects of mushroom production and industrial growth in India. Journal of Mushrooms. 2018;**16**(4):239-249

[108] Kazige OK, Chuma GB, Lusambya AS, Mondo JM, Balezi AZ. Valorizing staple crop residues through mushroom production to improve food security in eastern Democratic Republic of Congo. Journal of Agriculture and Food Research. 2022;**8**:100285

[109] Adjapong AO, Ansah KD, Angfaarabung F, Sintim HO. Maize residue as a viable substrate for farm scale cultivation of Oyster mushroom (Pleurotusosteatus). Advances in Agriculture. 2015;**2015**:1-6. DOI: 10.1155/2015/213251

[110] Royse DJ. Cultivation of Oyster Mushrooms. Extension Bulletin, College of Agricultural Sciences, The Pennsylvania State University, University Park, PA; 2003

[111] Wakchaure GC. Production and Marketing of Mushrooms: Global and national scenario. In: Singh M, Vijay B, Kamal S, Wakchaure GC, editors. Mushrooms Cultivation, Marketing and Consumption Global and National Scenario. Himachal Pradesh; 2011

[112] Polat E, Uzun H, Topçuo B, Önal K, Onus AN. Effects of spent mushroom compost on quality and productivity of cucumber (Cucumissativus L.) grown in greenhouses. African Journal of Biotechnology. 2009;**8**:176-180

[113] Nasehi M, Torbatinejad NM, Zerehdaran S. Effect of solid-state fermentation by oyster mushroom (Pleurotusflorida) on nutritive value of some agro by-products. Journal of Applied Animal Research. 2017;**45**:221-226

[114] Phan CW, Sabaratnam V. Potential uses of spent mushroom substrate and its associated lignocellulosic enzymes. Applied Microbiology and Biotechnology. 2012;**96**:863-873

[115] Appels FVW, Dijksterhuis J, LukasiewiczCE,JansenKMB,WöstenHAB, Krijgsheld P. Hydrophobin gene deletion and environmental growth conditions impact mechanical properties of mycelium by affecting the density of the material. Scientific Reports. 2018;**8**:4703

[116] Islam MR, Tudryn G, Bucinell R, Schadler L, Picu PC. Morphology and mechanics of fungal mycelium. Scientific Reports. 2017;**7**:13070

[117] Jones MP, Huynh T, Dekiwadia C, Daver F, John S. Mycelium composites: A review of engineering characteristics and growth kinetics. Journal of Bionanoscience. 2017;**11**:241-257

[118] Kulshreshtha S. Removal of pollutants using spent mushrooms substrates. Environmental Chemistry Letters. 2019;**17**:833-847

[119] Paula FS, Tatti E, Abram F, Wilson J, O'Flaherty V. Stabilisation of spent mushroom substrate for application as a plant growth-promoting organic amendment. Journal of Environmental Management. 2017;**196**:476-486

[120] Kumar K, Yadav AN, Kumar V, Vyas P, Dhaliwal HS. Food waste: A potential bioresource for extraction of nutraceuticals and bioactive compounds. Bioresources and Bioproceedings. 2017;**4**:1-14

[121] Rafael Gomes-Araújo R, Martínez-Vázquez DG, Charles-Rodríguez AV, Rangel-Ortega S, Robledo-Olivo A. Bioactive compounds from agricultural residues, their obtaining techniques, and the antimicrobial effect as postharvest additives. International Journal of Food Science. 2021;**2021**:1-13

[122] Jain S, Anal AK. Optimization of extraction of functional protein hydrolysates from chicken egg shell membrane (ESM) by ultrasonic assisted extraction (UAE) and enzymatic hydrolysis. Food Science and Technology. 2016;**69**:295-302. DOI: 10.1016/j.lwt.2016.01.057

[123] Maysonnave GS, Mello RO, Vaz FN, Avila MM, Pascoal LL, Rodrigues ACT.

Physicochemical characterization of by-products from beef cattle slaughter and economic feasibility of commercialization. Animal Science. 2020;**42**:e46545

[124] Prado DMF, Almeida AB, Oliveira-Filho JG, Alves CCF, Egea MB, Lemes AC. Extraction of bioactive proteins from seeds (Corn, Sorghum, and sunflower) and sunflower byproduct: Enzymatic hydrolysis and antioxidant properties. Current Nutrition & Food Science. 2020;**17**(3):310-320. DOI: 10.2174/1573401316999200731005803

[125] Kips L, De Paepe D, Bernaert N, Van Pamel E, De Loose M, Raes K, et al. Using a novel spiral-filter press technology to biorefine horticultural by-products: The case of tomato. Part I: Process optimization and evaluation of the process impact on the antioxidative capacity. Innovative Food Science & Emerging Technologies. 2016;**38**:198-205

[126] Alongi M, Anese M. Re-thinking functional food development through a holistic approach. Journal of Functional Foods. 2021;**81**:104466

[127] Silva LBAR, Pinheiro-Castro N, Novaes GM, et al. Bioactive food compounds, epigenetics and chronic disease prevention: Focus on early-life interventions with polyphenols. Food Research International. 2019;**125**:108646. DOI: 10.1016/j.foodres.2019.108646

[128] Valencia-Mejía E, Batista KA, Fernández JJA, Fernandes KF. Antihyperglycemic and hypoglycemic activity of naturally occurring peptides and protein hydrolysates from easy-to-cook and hard-to-cook beans (Phaseolus vulgaris L.). Food Research Interferon. 2019;**121**:238-246. DOI: 10.1016/j.foodres.2019.03.043

[129] Aguiar LM, Geraldi MV, Betim Cazarin CB, Marostica Junior MR. Chapter 11 - Functional food consumption and its physiological effects. In: Bioactive Compounds: Health Benefits and Potential Applications. M.R.S. Campos (editor) (Woodhead Publishing). 2019:205-225. DOI: 10.1016/b978-0-12-814774-0.00011-6

[130] Daliu P, Santini A, Novellino E. A Decade of neutraceutical patents:where are we now in 2018. Expert Opinion on Therapeutic Patents. 2018;**28**(12):875-882. DOI: 10.1080/13543776.2018.1552260

[131] Oliveira Filho JG, Braga ARC, de Oliveira BR, et al. The potential of anthocyanins in smart, active, and bioactive eco-friendly polymer-based films: A Review. Food Research International. 2021;**142**:10202

[132] Reque P, Brandelli A. Encapsulation of probiotics and nutraceuticals: Applications in functional food industry. Trends in Food Science and Technology. 2021;**114**:1-10. DOI: 10.1016/j.tifs.2021.05.022

[133] Ligianne Din Shirahigue LD, Ceccato-Antonini SR. Agro-industrial wastes as sources of bioactive compounds for food and fermentation industries. Ciência Rural, Santa Maria. 2020;**50**:e20190857

[134] Dini I. Bio discarded from waste to resource. Food. 2021;**10**:2652. DOI: 10.3390/foods10112652

[135] Martínez-Espinosa R. Introductory Chapter: A brief overview on fermentation and challenges for the next future. In: Martínez-Espinosa RM, editor. New Advances on Fermentation Processes. London, UK: IntechOpen; 2020

[136] Shin H-Y, Kim S-M, Lee JH, Lim S-T. Solid-state fermentation of black rice bran with Aspergillus awamori

and Aspergillus oryzae: Effects on phenolic acid composition and antioxidant activity of bran extracts. Food Chemistry. 2019;**272**:235-241. DOI: 10.1016/j.foodchem.2018.07.174

[137] Silva J, Honorato da Silva FL, Santos Ribeiro JE, Nóbrega de Melo DJ, Santos FA, Lucena de Medeiros L. Effect of supplementation, temperature and pH on carotenoids and lipids production by Rhodotorula mucilaginosa on Sisal bagasse hydrolyzate. Biocatalysis and Agricultural Biotechnology. 2020;**30**:101847. DOI: 10.1016/j.bcab.2020.101847

[138] Golebiewska E, Kalinowska M. Agricultural residues as a source of bioactive substances—waste management with the idea of circular economy. Environmental Science Proceedings. 2021;**9**(1):2

[139] Wenzel J, Samaniego CS, Wang L, Burrows L, Tucker E, Dwarshuis N, et al. Antioxidant potential of Juglans nigra, black walnut, husks extracted using supercritical carbon dioxide with an ethanol modifier. Food Science & Nutrition. 2016;**5**(2):223-232. DOI: 10.1002/fsn3.385

[140] Ko J-Y, Ko M-K, Kim D-S, Lim S-B. Enhanced production of phenolic compounds from pumpkin leaves by subcritical water hydrolysis. Preventive in Nutritional Food Science. 2016;**21**:132-137

[141] Mayanga-Torres PC, Lachos-Perez D, Rezende CA, Pradoc JM, Ma Z, Tompsett GT, et al. Valorization of coffee industry residues by subcritical water hydrolysis: Recovery of sugars and phenolic compounds. The Journal of Supercritical Fluids. 2017;**120**:75-85. DOI: 10.1016/j.supflu.2016.10.015

[142] Marathe SJ, Jadhav SB, Bankar SB, Dubey KK, Singhal RS. Improvements in the extraction of bioactive compounds by enzymes. Current Opinion in Food Science. 2019;**25**:62-72. DOI: 10.1016/j.cofs.2019.02.009

[143] Moreira MD, Melo MM, Coimbra JM, Reis KC, Schwan RF, Silva CF. Solid coffee waste as alternative to produce carotenoids with antioxidant and antimicrobial activities. Waste Management. 2018;**82**:93-99. DOI: 10.1016/j.wasman.2018.10.017

[144] Sinha S, Singh G, Arora A, Paul D. Carotenoid production by Red Yeast isolates grown in agricultural and "Mandi" waste. Waste and Biomass Valorization. 2021;**12**(7):3939-3949. DOI: 10.1007/s12649-020-01288-8

[145] Aguilo-Aguayo I, Walton J, Vinas I, Tiwari BK. Ultrasound assisted extraction of polysaccharides from mushroom by-products. LWT - Food Science and Technology. 2017;**77**:92-99

[146] Piñeiro Z, Marrufo-Curtido A, Serrano MJ, Palma M. Ultrasound-assisted extraction of stilbenes from grape canes. Molecules. 2016;**21**:784

[147] Zhou C, Wang Y. Recent progress in the conversion of biomass wastes into functional materials for value-added applications. Science and Technology of Advanced Materials. 2020;**21**(1):787-804. DOI: 10.1080/14686996.2020.1848213

[148] Filip S, Pavlic B, Vidovic S, Vladic J, Zekovic Z. Optimization of microwave-assisted extraction of polyphenolic compounds from Ocimum basilicum by response surface methodology. Food Analytical Methods. 2017a;**10**(7):2270-2280. DOI: 10.1007/s12161-017-0792-7

[149] Kulkarni V, Rathod V. Green process for extraction of mangiferin from Mangifera indica leaves. Journal of Biology. 2016;**6**:406-411

[150] Gencdag E, Gorguc A, Yilmaz FM. Recent advances in the recovery techniques of plant-based proteins from agro-industrial byproducts. Food Review International. 2021;**37**(4):447-468

[151] Isah S, Ozbay G. Valorization of food loss and wastes: Feedstocks for biofuels and valuable chemicals. Frontiers in Sustainable FoodSystem. 2020;**4**:82

[152] Mouratoglou E, Malliou V, Makris DP. Novel glycerol-based natural eutectic mixtures and their efficiency in the ultrasound-assisted extraction of antioxidant polyphenols from agri-food waste biomass. Waste and Biomass Valorization. 2016;**7**:1377-1387

[153] Sana B-O, Joudu I, Bhat R. Bioactives from agri-food wastes: Present insights and future challenges. Molecules. 2020;**25**:510. DOI: 10.3390/molecules25030510

[154] McClements DJ, Ozturk B. Utilization of nanotechnology to improve the application and bioavailability of phytochemicals derived from waste streams. Journal of Agricultural Food Chemistry. 2021;**70**(23):6884-6900

[155] Thakur J, Borah A. Microcapsules of bioactive compoundsfrom fruits and vegetables waste and their utilization: A review. Innovation. 2021;**10**(5):151-157

[156] Kennedy DO, Wightman EL. Herbal extracts and phytochemicals: Plant secondary metabolites and the enhancement of human brain function. Advance Nutrition. 2011;**2**:32-50. DOI: 10.3945/an.110.000117

[157] Friedman M. Analysis of biologically active compounds in potatoes (Solanum tuberosum), tomatoes (Lycopersiconesculentum), and jimson weed (Datura stramonium) seeds. Journal of Chromatography. A. 2004;**1054**:143-155

[158] Kalogeropoulus N, Chiou A, Pyriochou V, Peristeraki A, Karathanos KT. Bioactive phytochemicals in industrial tomatoes and their processing byproducts. LWT- Food Science and Technology. 2012;**49**(2):213-216

[159] Giovanetti MH, Nogueira A, de Oliveira CL, Wosiacki G. Chromatography—The Most Versatile Method of Chemical Analysis. Rijeka, Croatia: InTech; 2012

[160] Valdés A, Burgos N, Jiménez A, Garrigós MC. Natural pectin polysaccharides as edible coatings. Coatings. 2015;**5**:865-886

[161] Edris AE, Fadel HM. Investigation of the volatile aroma components of garlic leaves essential oil. Possibility of utilization to enrich garlic bulb oil. European Food Research and Technology. 2002;**214**:105-107

[162] BenítezV,MolláE,Martín-CabrejasMA, Aguilera Y, López-Andréu FJ, Cools K, et al. Characterization of industrial onion wastes (Allium cepa L.): Dietary fibre and bioactive compounds. Plant Foods for Human Nutrition. 2011;**66**:48-57

[163] Cáceres R, Malińska K, Marfà O. Nitrification within composting: A review. Waste Management. 2018;**72**:119-137

[164] Debode J, De Tender C, Cremelie P, Shein Lee A, Kyndt T, Muylle H, et al. Trichoderma-inoculated miscanthus straw can replace peat in strawberry cultivation, with beneficial effects on disease control. Frontiers in Plant Science. 2018;**9**:213

[165] Gervasi T, Pellizzeri V, Calabrese G, Di Bella G, Cicero N, Dugo G. Production of single cell protein (SCP) from food and agricultural waste by using Saccharomyces cerevisiae. Natural

Product Research. 2018;**32**(6):648-653. DOI: 10.1080/14786419.2017.1332617

[166] López-López N, López-Fabal A. Compost based ecological growing media according EU eco-label requirements. Scientia Horticulturae. 2016;**212**:1-10

[167] Rogers MA. Organic vegetable crop production in controlled environments using soilless media. Horticultural Technology. 2017;**27**:166-170

[168] Nayak A, Bhushan B. An overview of the recent trends on thewaste valorization techniques for food wastes. Journal of Environmental Management. 2019;*2019*(233):352-370

[169] Li Z, Song Z, Singh B, Wang H. The impact of crop residue biochars on silicon and nutrient cycles in croplands. Science Total Environment. 2019;**659**:673-680. DOI: 10.1016/j.scitotenv.2018.12.381

[170] Singh R, Babu J, Kumar R, Srivastava P, Singh P, Raghubanshi A. Multifaceted application of crop residue biochar as a tool for sustainable agriculture: An ecological perspective. Ecological Engineering. 2015;**77**:324-347. DOI: 10.1016/j. ecoleng.2015.01.011

[171] de la Rosa JM, Rosado M, Paneque M, Miller AZ, Knicker H. Effects of aging under field conditions on biochar structure and composition: Implications for biochar stability in soils. Science of the Total Environment. 2018;**613-614**:969-976. DOI: 10.1016/j.scitotenv.2017.09.124

[172] Katterer T, Roobroeck D, Andren O, Kimutai G, Karltun E, Kirchmann H, et al. Biochar addition persistently increased soil fertility and yields in maize-soybean rotations over 10 years in sub-humid regions of Kenya. Field Crops Research. 2019;**235**:18-26

[173] Roobroeck D, Hood-Nowotny R, Nakubulwa D, Tumuhairwe J-B. Biophysical potential of crop residues for biochar carbon sequestration, and co-benefits, in Uganda. Ecological Applications. 2019;**29**:E01984

[174] Minasny B, Malone BP, McBratney AB, Angeri DA, et al. Soil carbon 4 per mile. Geoderma. 2017;**292**:39-86

[175] Jeffery S, Abalos D, Prodana M, Bastos AC, Van Groenigen JW, Hungate BA, et al. Biochar boosts tropical but not temperate crop yields. Environmental Research Letters. 2017;**12**(5):053001. DOI: 10.1088/1748-9326/aa67bd

[176] Blanco-Canqui H. Biochar and soil physical properties. Soil Science Society of America Journal. 2017;**81**:687-711. DOI: 10.2136/sssaj2017.01.0017

[177] Yu L, Gamliel DP, Markunas B, Valla JA. A promising solution for food waste: Preparing activated carbons for phenol removal from water streams. ACS Omega. 2021;**6**:8870-8883

[178] Mohd Adib Yahya MA, Mansor MH, Zolkarnaini WAAW, Rusli NS, Aminuddin A, Mohamad K, et al. A brief review on activated carbon derived from agriculture by-Product. AIP Conference Proceedings. 2018;**1972**:030023

[179] Pradham BK, Sandle NK. Effect of different oxidizing agent treatments on the surface properties of carbon. Carbon. 1999;**4**:391-400

[180] Davis NE. Atherosclerosis-An inflammatory process. Journal of Insurance Medicine. 2005:**37**:72-75

[181] Majewska T, Pudyszak K, Koziowski K. The effect of charcoal addition to diets for broiler performance and carcass parameters. Veterinary Medicine Zootechnika. 2011;**55**(77):30-32

[182] Jiya EZ, Anyanwale BA, Ijaiya AT, Ugochukwu A, Tsado D. Effect of activated coconut shell charcoal meal on growth performance and nutrient digestibility of broiler chicken. British Journal of Applied Science and Technology. 2013;**3**(2):268-276

[183] Durunna CS, Abatai UG, Uchegbu C. Performance evaluation of broiler chickens fed varying levels of raw African velvet tamarind, Icheku (Dialium guineense) wood charcoal as feed additive. In: Proceedings of 43rdAnnual Conference of the Nigerian Society for Animal Production, March 18th -22nd 2018. Owerri, Imo State, Nigeria; 2018. pp. 514-516

[184] Dim CE, Akuru EA, Egom MA, Nnajofor NW, Ossai OK, Ukaigwe CG, et al. Effect of dietary inclusion of biochar on growth performance, haematology and serum lipid profile of broiler birds. Agro Science. 2018;**17**(2):8-16

[185] Linhoss JE, Purswell JL, Streetand JT, Rowland MR. Evaluation of biochar as a litter amendment for commercial broiler production. Journal of Applied Poultry Research. 2019;**28**:1089-1098

[186] Schmidt H-P, Hagemann N, Draper K, Kammann C. The use of biochar in animal feeding. PeerJ. 2019;7:e7373

[187] Zhang X, Wang H, He L, Lu K, Sarmah A, Li J, et al. Using biochar for remediation of soilscontaminated with heavy metals and organic pollutants. Environmental Science Pollution. 2013;**20**:8472-8483

[188] Moon DH, Hwang I, Chang YY, Koutsospyros A, Cheong KH, Ji WH, et al. Quality improvement of acidic soils by biochar derived from renewable materials. Environmental Science Pollution Controlled Series. 2017;**24**:4194-4199

[189] Zwieten LV, Kimber S, Morris S, Chan KY, Downie A, Rust J, et al. Effects of biochar from slow pyrolysis of papermill waste on agronomic performance and soil fertility. Plant Soil. 2010;**327**:235-246

[190] Chowdhury S, Sikder J, Mandal T, Halder G. Comprehensive analysis on sorptive uptake of enrofloxacin by activated carbon, and to enhance biodiesel production acting as a catalyst. Science Total Environment. 2019;**665**:438-452

[191] Dehkhoda AM, West AH, Ellis N. Biochar based solid acid catalyst for biodiesel production. Applied Catalysis A: General. 2010;**382**:197-204

[192] Yahya MA, Ngah CZC, Al-Qodah Z. Agricultural bio-waste materials as potential sustainable precursors used for activated carbon production: A review. Renewable and Sustainable Energy Reviews. 2015a;**46**:218-235. DOI: 10.1016/j.rser.2015.02.051

[193] Hameed BH, Daud FBM. Adsorption studies of basic dye on activated carbon derived from agricultural waste: *Hevea brasiliensis* seed coat. Chemical Engineering Journal. 2008;**139**(1):48-55

[194] Sugumaran PV, Susan P, Ravichandran P, Seshadri S. Production and characterization of activated carbon from banana empty fruit bunch and Delonix regia fruit pod. Journal of Sustainable Energy & Environment. 2012;**3**(3):125-132

[195] Amirza MAR, Adib MMR, Hamdan R. Application of agricultural wastes activated carbon for dye removal – an overview. MATEC Web of Conferences. 2017;**103**(06013):1-12

[196] Saygılı H, Güzel F. High surface area mesoporous activated carbon from tomato processing solid waste

by zinc chloride activation: Process optimization, characterization and dyes adsorption. Journal of Cleaner Production. 2016;**113**:995-1004

[197] Sych NV, Trofymenko SI, Poddubnaya OI, Tsyba MM, Sapsay VI, Klymchuk DO, et al. Porous structure and surface chemistry of phosphoric acid activated carbon from Corncob. Applied Surface Science. 2012;**261**:75-82

[198] Peng Y, Sun Y, Sun R, et al. Optimizing the synthesis of Fe/Al (Hydr)oxides-Biochars to maximize phosphate removal via response surface model. Journal of Cleaned Products. 2019;**237**:117770

[199] Ballu DA, Abdul-Halim A, Kojo QA, Yakubu A. Chemical characteristics of groundnut and sheanut shell biochars as adsorbents and soil conditioners in the era of ecological sustainability. Journal of Biological Research & Biotechnology. 2022;**20**(1):1461-1472

[200] Karri RR, Sahu JN. Modeling and optimization by particle swarm embedded neural network for adsorption of zinc (II) by Palm Kernel shell based activated carbon from aqueous environment. Journal of Environmental Management. 2018a;**206**:178-191

[201] Karri RR, Sahu JN. Process optimization and adsorptionmodeling using activated carbon derived from palm oil kernelshell for Zn (II) disposal from the aqueous environment usingdifferential evolution embedded neural network. Journal of Molecular Liquids. 2018b;**265**:592-602

[202] Hu Z, Srinivasan MP. Mesoporous high-surface-areaactivated carbon. Microporous and Mesoporous Materials. 2001;**43**:267-275

[203] Okhwa Hwang O, Lee S-R, Cho SS, Kyoung S, et al. Efficacy of different biochars in removing odorous volatile organic compounds (VOCs) emitted from swine manure. ACS Sustainable Chemical Engineering. 2018;**6**:14239-14247

[204] Niazi L, Lashanizadegan A, Sharififard H. Chestnut oakshells activated carbon: Preparation, characterization and application for Cr (VI) removal from dilute aqueous solutions. Journal of Cleaner Production. 2018;**185**:554-561

[205] Sathe PS, Adivarekar RV, Pandit AB. Valorization of peanut shell biochar for soil amendment. Journal of Plant Nutrition. 2021;**2021**:1-19. DOI: 10.1080/01904167.2021.1963771

[206] Sattar MS, Shakoor MB, Ali S, Rizwan M, Niazi NK, Jilani A. Comparativeefficiency of peanut shell and peanut shell biocharfor removal of arsenic from water. Environmental Science Pollution Research. 2019;**26**(18):18624-18635

[207] Tomul F, Yasin A, Burcu K, Trak D, Kendüzler E, Lima EC, et al. Peanut shells-derived biochars prepared from different carbonization processes: Comparison of characterization and mechanism of naproxen adsorption in water. Science of the Total Environment. 2020;**726**:137828. DOI: 10.1016/j.scitotenv.2020.137828

[208] Yu F, Sun L, Zhou Y, Gao B, Gao W, Bao C, et al. Biosorbents based on agricultural wastes for ionic liquid removal: An approach to agricultural wastes management. Chemosphere. 2016;**165**:94-99

[209] Karam DS, Nagabovanalli P, Rajoo KS, Ishak CF, Abdu A, Rosli Z, et al. An overview on the preparation of rice husk biochar, factors affecting its properties, and its agriculture application. Journal of the Saudi Society of Agricultural Sciences. 2022;**25**:149-159

[210] Phuong HT, Uddin M, Kato Y. Characterization of biochar from pyrolysis of rice husk and rice straw. Journal of Biobased Materials and Bioenergy. 2015;**9**:439-446

[211] Hesas RH, Arami-Niya A, Wan Daud WMA, Sahu JN. Preparation and characterization of activated carbon from applewaste by microwave-assisted phosphoric acid activation: Application in methylene blue adsorption. BioResources. 2013;**8**:2950-2966

[212] Ozdemir I, Şahin M, Orhan R, Erdem. Preparation and characterization of activated carbon from grape stalk by zinc chloride activation. Fuel Processing Technology. 2014;**125**:200-206

[213] Ren J, Chen N, Wan L, Li G, Chen T, Yang F, et al. Preparation of high-performance activated carbon from coffee grounds after extraction of bio-oil. Molecules. 2021;**26**:257

[214] Promraksa A, Rakmak N. Biochar production from palm oil mill residues and application of the biochar to adsorb carbon dioxide. Heliyon. 2020;**6**:e04019. DOI: 10.1016/j.heliyon.2020.e04019

[215] Safana Aminu Aliyu SA, Abdullah N, Sulaiman F, Umar S. Combustion characteristics of palm pressed fibres biochar and subbituminous Malaysian coal. Malaysian Journal of Fundamental and Applied Sciences. 2018;**14**(3):334-337

[216] Adilah Shariff A, Aziz NSM, Abdullah N. Slow pyrolysis of oil palm empty fruit bunches for biochar production and characterization. Journal of Physical Science. 2014;**25**(2):97-112

[217] Rahayu DE, Karnaningroem N, Altway A, Slamet A. Utilization of oil palm empty fruit bunches biomass through slow pyrolysis process. IOP Conference Series. 2021;**913**:012018

[218] Yasin J, Pravinkumar R. Production of activated carbon from bio-waste materials by chemical activation method. AIP Conference Proceedings. 2020;**2225**:070005

[219] Chaves L, Fernandes JD, Mendes JS, Dantas ERB, Guerra HC, Tito GA, et al. Characterization of poultry litter biochar for agricultural use. SYLWAN. 2020;**164**(6):468-487

[220] Shakya A, Agarwal T. Poultry litter biochar: An approach towards poultry litter management – a review. International Journal of Current Microbiology. 2017;**6**(10):2657-2668

[221] Zhou Y, Lu J, Zhou Y, Liu Y. Recent advances for dyes removal using novel adsorbents: A review. Environmental Pollution. 2019;**252**:352-365

[222] Ma YL, Matsunaka T. Biochar derived from dairy cattle carcasses as an alternative source of phosphorus and amendment for soil acidity. Soil Science and Plant Nutrition. 2013;**59**(4):628-641. DOI: 10.1080/00380768.2013.806205

[223] Gunamantha IM, Widana GAB. Characterization the potential of biochar from cow and pig manure for geoecology application. IOP Conference Series. 2018;**2018**:131012055

[224] Chen HY, Li WY, Wang JJ, Xu HJ, Liu YL, Zhang Z, et al. Adsorption of cadmium and lead ions by phosphoric acid-modified biochar generated from chicken feather: Selective adsorption and influence of dissolved organic matter. Bioresource Technology. 2019;**292**:121948

[225] Zhang C, Alexis WD, Li H, Guo M. Use of poultry litter derived biochar as litter amendment to control ammonia emissions. In: ASABE Annual International Meeting. Orlando, FL; 2016a

[226] Zdravkov BD, Čermák JJ, Šefara M, Janku J. Pore classification in

the characterization of porous materials: A perspective. Central European Journal of Chemistry. 2007;**5**(**2**):385-395. DOI: 10.2478/s11532-007-0017-9

[227] Marsh H, Rodriguez-Reinoso F. Activated Carbon. 1st ed. New York: Elsevier Science and Technology Books; 2006

[228] Budi E, Nasbey H, Yuniarti BDP, Nurmayatri Y, Fahdiana J, Budi AS. Pore structure of the activated coconut shell charcoal carbon. AIP Conference Proceedings. 2014;**1617**:130. DOI: 10.1063/1.4897121

[229] Bouchelta C, Medjram MS, Bertrand O, Bellat J-P. Preparation and characterization of activated carbon from datestones by physical activation with steam. Journal of Analytical and Applied Pyrolysis. 2008;**82**:70-77

[230] González JF, Román S, Encinar JM, Martínez G. Pyrolysis of various biomass residues and char utilization for theproduction of activated carbons. Journal of Analytical Applied Pyrolysis. 2009;**85**:134-141

[231] Esmar Budi E, Umiatin H, Nasbey RA, Bintoro F. Activated coconut shell charcoal carbon using chemicalphysical activation. AIP Conference Proceedings. 2016;**1712**:050003. DOI: 10.1063/1.4941886

[232] Adinata D, Wan Daud WMA, Aroua MK. Preparationand characterization of activated carbon from palm shell bychemical activation with K2CO3. Bioresource Technology. 2007;**98**:145-149

[233] Tseng R-L. Physical and chemical properties and adsorptiontype of activated carbon prepared from plum kernels by NaOHactivation. Journal of Hazardous Materials. 2007;**147**:1020-1027

[234] Danish M, Hashim R, Ibrahim MNM, Sulaiman O. Effect of acidic activating agents on surface area and surface functional groups of activated carbons produced from Acaciamangium wood. Journal of Analytical and Applied Pyrolysis. 2013;**104**:418-425

[235] Filip S, Pavlić B, Vidović S, Vladić J, Zeković Z. Optimization of microwave-assisted extraction of polyphenolic compounds from Ocimum basilicum by response surface methodology. Food Analytical Methods. 2017b. DOI: 10.1007/s12161-017-0792-7

[236] Karagoz S, Tay T, Ucar S, Erdem M. Activated carbonsfrom waste biomass by sulfuric acid activation and their use onmethylene blue adsorption. Bioresource Technology. 2008;**99**:6214-6222

[237] Kyzas GZ, Deliyanni EA, Matis KA. Activated carbons produced by pyrolysis of waste potato peels: Cobalt ions removal by adsorption. Colloids and Surfaces, A: Physicochemical and Engineering Aspects. 2016;**490**:74-83

[238] Yahya MA, Al-Qodah Z, Ngah CWZ. Agricultural biowaste materials as potential sustainable precursors used for activated carbon production: A review. Renewable Sustainable Energy Review. 2015b;**46**:218-235

[239] Zhang S, Tao L, Zhang Y, Wang Z, Gou G, Jiang M, et al. Powder Technology. 2016b;**295**:152-160

Chapter 6

Crop Residue Burning in India: Potential Solutions

Kawaljeet Kaur and Preetpal Singh

Abstract

With its second-largest agro-based economy and year-round crop production, India produces a lot of agricultural waste, including crop residues. Because India lacks effective sustainable management methods, an estimated 92 seems like a very small quantity of metric tons of crop waste burned each year, causing excessive particulate matter emissions and air pollution. Burning crop residue has grown into a serious environmental problem that threatens human health and causes global warming. Composting, making biochar, and mechanization are a few effective sustainable solutions that can assist in resolving the issue while maintaining the nutrients found in the agricultural residue in the soil. In order to promote environmentally friendly management practices, the Indian government has launched a number of programs and campaigns.

Keywords: India, agricultural waste, crop residue, field residue, process residue, crop residue burning, biochar, composting, biogas, policy challenges

1. Introduction

The global economic expansion is significantly influenced by the agricultural sector. However, the handling of agricultural waste receives scant attention in the literature. It might be connected to the fact that the agricultural industry is not as strictly controlled as the municipal solid waste industry (MSW). Since municipalities and other public institutions are primarily in charge of managing MSW, data on generation and management are gathered, kept track of, and examined in public. Agricultural wastes are materials left over after different agricultural processes. Agricultural waste, according to the United Nations, often consists of manure and other wastes from farms, poultry houses, and slaughterhouses; harvest waste; fertilizer run-off from fields; pesticides that enter water, air, or soils; and salt and silt drained from fields [1–3]. With little engagement from the public sector, agricultural waste is primarily managed by the owners of the agricultural land, who are primarily in the private sector. The world's food output has greatly increased as a result of the rising food demand in developing nations. The multiplicity of agricultural operations raises the quantity of agricultural products produced, which has an overall negative impact on the environment by increasing waste production. Due to advancements in water management systems, contemporary agro technologies, and extensive pesticide deployment, enormous swaths of wasteland have been transformed into agricultural fields [1]. These actions have exacerbated environmental degradation on a global scale and complicated

IntechOpen

Figure 1.
Burning of rice residues, a prevalent practice in northwest India.

the process of disposing of agricultural waste. To manage these wastes, including their conversion into useable resources, the national agencies are continuously creating policies and other potential approaches. The term "harvest trash," more often known as "crop residue," refers to both the field residues that remain in an agricultural field or orchard after the crop is harvested and the process residues that are left over after the crop is processed into a useful resource. Field remnants commonly include stalks and stubble (stems), leaves, and seedpods. Molasses and sugarcane bagasse are two examples of process leftovers that are useful (**Figure 1**) [2, 4, 5].

2. Crop residue: composition and decomposing mechanisms

These crop residues are a summary of the general categories of crop residues produced by the major cereal crops and sugar cane, in particular as a field residue is a natural resource that traditionally contributed to the soil stability and fertility through direct plowing into the soil or by composting. Cellulose, hemicellulose, and lignin make up the majority of plant biomass, with smaller amounts of pectin, protein extractives, sugars, nitrogenous material, chlorophyll, and inorganic waste [6–8]. Lignin is almost impermeable and offers the structural support compared with cellulose and hemicellulose. Lignin is extremely resistant to both chemical and biological degradation, which helps it resist fermentation [8, 9]. The term "lignocellulosic biomass" refers to the parts of plants that are not used for food, such as the stalks, straw, and husk [5]. The majority of the lignocellulosic biomass is made up of the four most important agricultural crops farmed worldwide: sugarcane, wheat, rice, and maize. The lignin layer is typically pretreated with lignin degrading microorganisms to break down the lignin layer and degrade cellulose and hemicellulose matter to the corresponding monomers and sugars for efficient biomass to fuel conversion

[6] because it is resistant to chemical and biological degradation by fungi, bacteria, and enzymes. Mechanical, chemical, physicochemical, or biological pretreatment options are available. These techniques lead to an increase in the accessible surface area, porosity, and degree of polymerization, as well as a decrease in the crystallinity of cellulose and hemicellulose. Utilizing microorganisms to control agricultural waste could also be a great way to detoxify the soil and reduce environmental pollution [10]. The complex materials in the biomass are broken down by microbial communities into simpler elements that can be recycled or reused in other parts of the ecosystem. Depending on the type of bacteria, fungi, or algae involved in the degradation, the processes used can either be aerobic or anaerobic [11, 12].

3. Adverse impact of crop residue burning on the environment

Crop residue burning causes a variety of environmental issues. Burning crop residue has several negative effects, but the main ones are the release of greenhouse gases (GHGs) that contribute to global warming, elevated levels of particulate matter (PM), and smog that pose health risks, loss of agricultural lands' biodiversity, and deterioration of soil fertility [13]. Burning crop residue dramatically raises air pollution levels of CO_2, CO, NH_3, NOX, SOX, non-methane hydrocarbon (NMHC), volatile organic compounds (VOCs), semivolatile organic compounds (SVOCs), and particulate matter (PM) [14, 15]. In essence, this explains why organic carbon, nitrogen, and other nutrients that would normally have been kept in the soil have been lost [13, 16].

Crop residue burning in Delhi produces 17 times as much particulate matter (PM) as all other sources combined, including industry, burning of waste, and vehicle emissions [17]. As a result, the residue burning in India's northwest produces almost 20% of the country's total organic and elemental carbon emissions from burning agricultural waste [13]. Crop burning contributes greatly to climate change by raising the amount of particulates (PM) in the atmosphere. The release of fine black and brown carbon (primary and secondary), which alters light absorption, is one factor in global climate change [7, 18, 19]. According to their aerodynamic diameter and chemical makeup, PM2.5 and PM10 particles in the air are typically divided into two categories: fine and coarse particles, respectively. PM2.5 particles have an aerodynamic diameter of less than 2.5 and 10 m, respectively. Lightweight particles can move farther with the wind and can remain suspended in the air for longer [16, 20]. Because the particles are light and linger in the air for a longer period of time than heavier ones do, the effect of particulate matter is exacerbated by meteorological conditions. In the Patiala area of Punjab, the yearly contribution of PM2.5 from paddy residue burning was estimated to be between 60 and 390 mg/m^3 [13].

4. Sustainable management practices for crop residue

Over the past 10 years, scientists and agriculturalists have long recommended alternative strategies to prevent crop residue burning, but due to farmers' lack of understanding and social conscience, these measures have not been properly adopted. This section contains information on three such agricultural applications that have, for a variety of reasons, either been disregarded or skipped. They are in-situ management by mechanical intensification, biochar, and composting.

5. Composting

Composting is the naturally occurring, under regulated conditions, rotting or breakdown of organic waste by microorganisms [21]. Compost, which is a rich source of organic matter, is crucial for maintaining soil fertility and promoting sustainable agricultural output. Composting the soil enhances its physical, chemical, and biological qualities and can entirely replace the use of agricultural chemicals such as fertilizer and pesticides. The advantageous impacts of compost supplemented soil include greater potential for increased yields and resilience to environmental conditions such as drought, disease, and toxicity [21–23]. Due to increased soil microbial activity, these methods also aid in greater nutrient uptake and active nutrient cycling.

The organic matter is treated twice during the composting process.

i. **Degradation:** The initial stage of degradation begins with the breakdown of organic compounds that are simple to digest, such as sugars, amino acids, and organic acids. In addition to releasing carbon dioxide and energy, aerobic microbes also absorb oxygen. The initial thermophilic phase, which lasts for a few weeks to months, is characterized by high temperature, high pH, and humidity, all of which are necessary for activating the microorganisms [24]. Additionally, it is made sure that the substrate is adequately supplied with oxygen during this period [25].

ii. **Maturation:** During the subsequent few weeks, more complex organic compounds are broken down, which is followed by a decline in the microbial population. When the temperature drops to 40–45°C, the phase transition from thermophilic to mesophilic occurs [25–27]. At the last step, the system's biological activity decreases as the temperature falls to an ambient level. Finally, a soil-like substance with a dark brown to black hue is created. This soil-like substance also has a higher humus content, a lower carbon-nitrogen ratio, and a pH that has been neutralized [21]. Eventually, the biomass is changed into a nutrient-rich substance that can enhance the soil's structural qualities [28].

6. Biochar

The thermochemical process known as pyrolysis, which occurs at low temperatures in an oxygen-free atmosphere, produces biochar, a porous material with fine-grained carbon content [29]. It is a mixture of varying amounts of carbon (C), hydrogen (H), oxygen (O), nitrogen (N), sulphur (S), and ash [30]. The very porous characteristic of biochar, when added to soil, aids in better water retention and increased soil surface area. As a result, there is more interest in using soil amendments such as biochar, black carbon, and charcoal to stabilize soil organic content. These methods are thought to be an effective way to reduce agricultural waste while also reducing GHG emissions. In order to reduce the amount of CO_2 or methane released into the atmosphere, the process of carbon sequestration essentially calls for longer residence times and resistance to chemical oxidation of biomass to CO_2 or reduction to methane [8, 9]. The partially burned byproducts are pyrogenic carbon/carbon black, which undergoes a very slow chemical change to create a long-term carbon sink that is perfect for soil amendment [31, 32]. It primarily interacts with soil bacteria, plant roots, and the soil matrix [33]. It also aids in nutrient retention and triggers a variety of biogeochemical processes. Currently, biochar is being used sparingly in India, mostly in villages and small towns. Promoting the biochar process in India would be more advantageous given its broad applicability.

7. In-situ management with mechanical intensification

Many farmers use in-situ application of crop residue because it is a natural process. This process also gives the soil certain advantages. There are two primary techniques for applying chemicals in the field, but both entail leaving crop residue on the fields after harvest. What occurs with tillage in the following season will determine how they differ. In the first approach, planting is done the next season with little to no tillage, whereas in the second way, crop residue is mechanically absorbed into the soil during plowing [34]. Both techniques require specialized (new) equipment, such as machinery for crop residue absorption into soils or no-till seeing equipment, even though in-situ management of agricultural residues can offer long-term cost savings on equipment and manpower. In North America, crop residue retention with no-tillage is primarily used, and in the United States alone, no-till farming accounts for about 40% of cropland [30]. In-situ management techniques such as direct incorporation into soils and mulching are specifically mentioned in the National Policy for Management of Crop Residue [13] as ones that should be promoted in India not only to control crop residue burning but also to prevent environmental degradation in the croplands.

8. Soil health and conservation agriculture

A soil must, among other things, have room for plant roots to spread out, be able to store and make water and nutrients available to plant roots, and offer a favorable biotic and chemical environment for soil microorganisms to function in order to maintain soil porosity, fix atmospheric nitrogen, hold, and mineralize nutrients. These factors must work in concert to create the foundation for the defined soil health.

With the help of beneficial symbiotic relationships with plant roots, such as those formed by nitrogen-fixing bacteria and mycorrhizal fungi, recycling vital plant nutrients, and improvements to soil structure (such as aggregate stability), which in turn improves soil water and nutrient holding capacity, healthy soils maintain a diverse community of soil organisms that help to control plant disease, insect, and weed pests, and ultimately improve crop production. Many regions of the world agree that their soils are ill, unhealthy, and lacking in the ability to produce enough food for themselves. While "soil quality" is frequently mentioned as if it were a fixed quality, "soil health" is less frequently mentioned and refers specifically to the biological dynamics of soil quality.

- The following essential CA elements benefit soil in "good condition" (static) or "good health" (dynamic):

- Minimal disruption of the ideal porous soil architecture, which (a) maintains optimal levels of respiration gases in the rooting zone; (b) moderates oxidation of organic matter; (c) facilitates water movement, retention, and release at all scales; and (d) restricts re-exposure of weed seeds and their germination.

- The soil surface benefits from a permanent layer of sufficient organic matter, particularly crop residues, including: (a) protection from the harsh effects of solar radiation and rain; (b) a substrate for soil organism activity; (c) increased cation-exchange capacity for nutrient capture, retention, and slow release; and (d) weed smothering. Legumes included in crop rotations and sequences offer the following benefits:

- A range of species, for direct harvest and/or fodder; (a) minimal rates of pest species population build-up through life-cycle disruption; (b) biological N-fixation in suitable conditions, limiting external costs; (c) prolonged slow-release of such N from complex organic molecules derived from soil organisms; and (e) improvement of soil profile by organic matter addition at all depths.

According to the Ministry of New and Renewable Energy (MNRE 2009), the government of India, approximately 500 Mt. of crop residue is produced annually. Depending on the cropping intensity, productivity, and crops planted in various Indian states, there is a wide variation in crop residue generation and utilization. The most waste is produced in Uttar Pradesh (60 Mt), then in Punjab (51 Mt), and then in Maharashtra (46 Mt). Cereals provide 352 Mt. of leftovers from various crops, which is followed by fibers (66 Mt), oilseeds (29 Mt), pulses (13 Mt), and sugarcane (12 Mt). The cereal crops (rice, wheat, maize, and millets) account for 70% of crop residues, whereas the rice crop alone accounts for 34%. With 22% of the residues produced, wheat comes in second, while 13% of the residues produced by all crops come from fiber crops. Cotton produces the most fiber (53 Mt) and has an 11% crop residual rate. Coconut comes in second among fiber crops for residue generation with 12 Mt. In India, crop residues made up of tops and leaves from sugarcane production total 12 Mt., or 2% of all crop residues.

The excess leftovers, or those that were generated but not used for other purposes, are often burned in the field or used to power homes. India's estimated annual crop residual surplus ranges from 84 to 141 Mt., with grains and fiber crops accounting for 58 and 23%, respectively, of the total. The remaining 19% comes from various crops, sugarcane, legumes, and oilseeds. Out of the 82 Mt. of excess cereal crop residues, 44 Mt. are leftover rice, followed by 24.5 Mt. of wheat, which is primarily burned in fields. An estimated 80% of the extra residue from fiber crops (33 Mt) is cotton, and this residue is burned.

9. Remainders' positive influence on soil health

Crop residues can have a number of beneficial effects on the physical, chemical, and biological aspects of soil, whether they are incorporated into the soil or are left on the surface. By changing the soil's structure and aggregate stability, it lowers the bulk density of the soil and increases hydraulic conductivity. Plant residue mulching increases the minimum soil temperature in winter owing to a decrease in the upward heat flux from the soil and lowers the minimum soil temperature in summer due to the shade effect. Crop residues that are kept on the soil's surface slow runoff by acting as miniscule dams, prevent the formation of surface crusts, and improve infiltration. When left unaltered with no-till, the channels (macro-pores) made by earthworms and old plant roots enhance infiltration to aid in reducing or eliminating runoff. A higher level of soil moisture can, in many cropping and climatic settings, lead to a higher crop yield when combined with decreased water evaporation from the top few inches of soil and improved soil properties.

Residues serve as a store for plant nutrients, stop nutrient leaching, boost cation exchange capacity (CEC), offer a hospitable environment for biological N fixation, boost microbial biomass, and improve the activities of enzymes such as dehydrogenase and alkaline phosphatase. Increased microbial biomass can improve soil nutrient availability and serve as a source and sink of nutrients for plants. The reduction of wind and water erosion, the improvement of water infiltration and moisture retention, and the reduction of surface sediment and water runoff are all benefits of leaving significant amounts of crop residues equally dispersed across the soil surface. Crop residues are crucial in reducing soil acidity by releasing hydroxyls, especially during the breakdown of residues with higher C:N ratios, and in increasing soil alkalinity by applying residues with lower C:N ratios.crops including legumes: oilseeds and pulses. Crop residues' contribution to soil carbon sequestration would be a bonus in terms of managing the effects of climate change.

10. Use of residues in conservation agriculture is subject to restrictions

With greater residue levels in Conservation Agriculture, there are a number of difficulties (CA). These include issues with various diseases, insects, or weeds as well as challenges caused by increased surface residues to effective seed, fertilizer, and pesticide placement. With their greater amounts of crop residue, conservation tillage approaches typically demand more care, timing, placement of nutrients and pesticides, and tillage operations. Due to increasing residue levels and fewer options for the manner and timing of nutrient administrations, nutrient management may become problematic. In instance, no-till can make it more difficult to apply manure and may cause nutrient stratification in the soil profile as a result of repeated surface treatments without any mechanical assimilation.

Placement of seed at the right depth to promote germination in the no-tilled plots with residue kept on the soil surface is still a challenge and is one of the major technological bottlenecks that requires attention. Although the zero-till seed-cum-fertilizer drill machinery has undergone significant progress, there is still more room for advancement to provide farmers with a hassle-free technology. The other bottleneck, particularly in the rice-wheat system, is weed control. Given that chemical herbicides can leak into the environment, overuse of them may not be ideal. All fertilizers, notably N, should not be applied as a base dose at the time of planting because doing so could reduce their effectiveness and pollute the environment. Although evaporation is decreased and more water is kept close to the top with greater residue levels, this stimulates the establishment of feeder roots close to the surface where the nutrients are concentrated. Higher expenditures may occasionally result from additional application of particular nutrients and the need for specialized equipment for efficient fertilizer placement. Similar to how higher pesticide use may be required for CA adoption. The problem of non-point source pollution and environmental hazard is already present in the nations that use proportionally more herbicides.

- Additional management skill requirements, concerns about poorer crop yields and/or economic returns, unfavorable attitudes or views, and institutional restrictions are additional barriers to farmers adopting residue integration systems. Farmers and occasionally entire communities show a great preference for well-kept tilled fields. They take great satisfaction in keeping their fields "clean" of debris and actively tilling them to create a flat surface before planting.

- Other applications for crop leftovers

- There are a number of strategies that can be used to handle residues effectively. Large quantities of wastes can be utilized for compost preparation, energy production, the creation of biofuel, and mushroom culture in addition to being used as cattle feed.

By utilizing the leftovers as animal bedding and subsequently piling them in a dung pit, the wastes can be composted. A kilogram of straw can hold up to 2–3 kilogram of animal excreted pee. On the farm itself, material can also be composted using various techniques. One hectare's worth of rice leftover produces 3.2 tonnes of nutrient-rich farmyard manure (FYM).

11. Biomass energy from crop waste

Because of its benefits for the environment, biomass is a source of energy that can be used effectively and is sought for on a global scale. Crop residue is now being used more frequently to produce energy and to replace fossil fuels in recent years.

Additionally, it provides a quick fix for lowering the atmospheric CO_2 concentration. Biomass is a storable resource that is less expensive, more energy-efficient, and environmentally friendlier than other renewable energy sources such as solar and wind energy. Straw, however, has a low bulk density and a low energy output on a weight basis. Regardless of the bio-energy technology, the logistics of delivering the vast quantities of straw needed for efficient energy generation is a significant economic factor. The capacity to use residues for energy generation is influenced by a number of factors, including the availability of residues, the cost of transportation, and the infrastructure (such as harvesting equipment and collection methods).

12. Ethanol produced from crop waste

Because ethanol can be used as a pure fuel in internal combustion engines or combined with gasoline as a fuel extender and octane-enhancing agent, the conversion of ligno-cellulosic biomass into bio-based alcohol production is a significant and researchable subject. Theoretical estimates of the amount of ethanol that may be produced from various feedstocks, including maize grain, wheat straw, rice straw, bagasse, and sawdust, range from 382 to 471 $L\ t^{-1}$ of dry matter.

13. Biomethanation

Biogas, a gaseous mixture of carbon dioxide and methane, can be produced from biomass, such as rice straw, and used as fuel. It has been claimed that dry rice straw may produce 300 $m^3\ t^{-1}$ of biogas. The procedure produces good-quality gas with a methane content of 55–60%, and the spent slurry can be used as manure. This approach aims to produce manure that may be returned into the soil while also using agricultural waste to extract high-quality fuel gas.

14. Gasification of waste

Gas is created during the thermochemical process of gasification when wastes partially burn. After early pyrolysis, the process thoroughly decomposes the biomass to produce energy-rich gaseous products. The primary issue in using biomass gasification to produce electricity is purifying the gas to get rid of pollutants. The residues can be put to use in gasifiers to create producer gas. In some states, gasifiers with a capacity of more than 1 MW have been erected to produce producer gas, which is fed to motors connected to alternators to produce power. About 300 kWh of power may be produced from 1 tonne of biomass.

15. Fast pyrolysis

In order to quickly pyrolyze crop leftovers, the biomass must quickly reach a temperature of 400–500°C. This causes a striking modification in the thermal disintegration process. A biomass's dry weight is transformed into condensable vapors to the tune of about 75%. The condensate produces a dark brown, viscous liquid known as bio-oil if it cools down quickly within a few seconds. Bio-oil has a calorific value that ranges from 16 to 20 MJ kg^{-1}.

16. Biochar

Biochar is a high-carbon substance made from biomass that has been slowly heated without oxygen. It has benefits in terms of its effectiveness as an energy source, its usage as a fertilizer when combined with soil, and its capacity to stabilize and lower atmospheric emissions of hazardous gases. Biochar is useful for releasing gases that are high in energy, which are subsequently used to create liquid fuels or directly to generate electricity and/or heat. It might have a significant impact on the long-term storage of carbon. Biochar improves the soil's fertility and capacity to hold onto water, as well as speeding up the transport of minerals to plant roots.

17. Elements of a nutrient management strategy in CA

Components of a nutrient management plan in California

As an agro-ecological technique with a biological foundation, CA does not concentrate on a specific product or species. Instead, it deals with the intricate relationships between various crops and specific local conditions, capitalizing on the intricate networks of relationships involved in managing soil systems in a profitable and sustainable manner.

The following four general aspects would need to be taken into account by nutrient management techniques in CA systems:

I. The biological processes of the soil are strengthened and safeguarded to provide all soil biota and microorganisms preference and to increase and maintain soil organic matter and soil porosity;

II. Enough biomass is produced and biological nitrogen is fixed to maintain enough soil energy and nutrient stores.

III. The production of biomass and biological nitrogen fixation is sufficient to maintain the energy and nutrient levels in the soil necessary to support higher levels of biological activity and to cover the soil;

IV. Plant roots in the soil have sufficient access to all nutrients from both natural and artificial sources to meet crop needs; and.

V. The pH of the soil is maintained within an appropriate range to ensure that all important soil chemical and biological processes run smoothly.

18. Managing soil biological processes: the living system that is soil

Water must enter soil that is permeable from the surface below in order for rivers, plants, and groundwater to function. The interaction between the biological, physical, and chemical aspects of soil productivity is hampered by a lack of water for plants. The volume and interconnectedness of soil pores, which in turn control their ability to transfer water, determine the rate at which water enters, passes through, and moves through the soil.

The percentage of soil pores that can hold water against gravity and still release it in reaction to "suction" produced by roots, as defined by the physiology of the plants and atmospheric demand, determines the volume and availability of water that plants may use. Nutrient management and water management in soil are inextricably related.

The productivity of the soil in which plants are growing is reduced due to a lack of water and/or other nutrients, which prevents the complete interaction of the plant and soil systems. Plant growth and development are hampered by inadequate plant nutrients, and a severe water shortage brings the system to a complete stop.

Due to degradation and loss of organic matter, soil porosity is compromised or eliminated through compaction, pulverization, and/or collapse.

Due to the deterioration and loss of organic matter, compaction, pulverization, and/or collapse harm or destroy soil porosity. Tillage of the soil causes a rapid oxidation of the organic carbon in the materials to carbon dioxide gas and its loss to the environment, resulting in a net loss of organic matter. After such damage, the nonliving portion of organic matter is biologically transformed by its living component, the fauna and flora that live the soil, ranging from microorganisms such as bacteria to macroorganisms such as worms, termites, and the plants themselves. The creation of irregular aggregates of soil particles, within and between which are the crucial pore-spaces in soil, is facilitated by their metabolic activity, which also produces glue-like compounds, fungal hyphen, etc. The production of irregular aggregates of soil particles, within and between, which are the vital pore-spaces in which water, oxygen, and carbon dioxide move and roots grow, is a result of their metabolic activity, which also contributes glue-like compounds, fungal filaments, etc. Additionally, these elements significantly increase the soil's ability to grab and hold onto nutrient ions on organic complexes and provide a slow-release mechanism for their release back into the soil's moisture. A significant amount of fresh organic matter must constantly be present in the soil as a source of energy and nutrients for soil organisms, not simply for the plants alone, in order for this activity and its effects to cease. If the conditions for biotic activity in the soil are kept favorable, this dynamic process of formation and reformation of the porous soil architecture will continue from year to year, maintaining the capacities of landscapes thus treated to continue yielding vegetation and water on a regular basis, contributing to the sustainability of such production processes.

This is where maintaining "soil health" becomes important. It is more appropriate to think of the soil as primarily a living, porous biological entity that penetrates the nonliving components and forms from the top downward, rather than as a geological entity that forms from the bottom upwards with living things in it at the top, when deciding how to manage the land and nutrients to maintain its productivity [25].

19. Government intervention

The engagement of the right government agencies is required to implement strict measures to reduce crop burning and better control crop waste management. The Indian government has made several attempts to introduce and educate the agricultural community about best practices. Environmentalists and government officials have also developed ideas to reduce crop residue burning and encourage the use of alternative sustainable management techniques. The National Environment Appellate Authority Act of 1997, the Air Prevention and Control of Pollution Act of 1981, the Environment Protection Act of 1986, the National Tribunal Act of 1995, and Section 144 of the Civil Procedure Code (CPC) are a few of the laws relating to crop residue burning that are currently in effect. The National Green Tribunal (NGT) has implemented strict regulations, particularly in the states of Rajasthan, Uttar Pradesh, Haryana, and Punjab, to reduce crop residue burning [13, 35].

20. National schemes and policies

The National Thermal Power Corporation (NTPC) has lately been instructed by the Indian government to combine crop residue pellets (almost 10%) with coal for the purpose of generating electricity [36]. The farmers benefited from a financial return of about Rs. 5500 (77 USD) per tonne of agricultural leftovers as a result. These profitable techniques have not yet been implemented, but farmers might profitably take advantage of them. The Indian government only operates a few bio-composting-related measures. As a part of its 11th Five Year Plan, the Indian government announced the Rashtriya Krishi Vikas Yogna (RKVY), State Plan Scheme of Additional Central Assistance, in August 2007 (**Table 1**) [35].

Query	Response	Crop residues management options
1. Can crop residues be used for conservation agriculture?	Yes	• Retain it on soil surface • Use drill for sowing with residues (e.g. Happy Seeder)
If the answer is No, move to query 2		• Follow conservation agriculture for all crops in rotation
2. Can it be used as fodder?	Yes	• Leave stubbles in field • Enrich fodder with
If the answer is No, move to query 3		supplements (e.g. urea and molasses) • Use manure in conservation agriculture
3. Can it be used for biogas generation?	Yes	• Leave stubbles in field • Adopt community biogas
If the answer is No, move to query 4		plant (e.g. KVIC design modified by IARI) • Use slurry in conservation agriculture
Query	Response	Crop residues management options
4. Can it be used for composting?	Yes	• Leave stubbles in field • Adopt modern composting technique (e.g. IARI model)
If the answer is No, move to query 5		• Use compost in conservation agriculture
5. Can it be used for bio-fuel generation?	Yes	• Leave stubbles in field • Install bio-fuel plant • Use liquid slurry in conservation agriculture
If the answer is No, move to query 6		
6. Can it be used for electricity generation?	Yes	• Leave stubbles in field • Install biomass-energy plant (e.g. KPTL model)
If the answer is No, move to query 7		• Use ash in conservation agriculture
7. Can it be used for gasification?	Yes	• Leave stubbles in field • Install biomass gasifier (e.g. CIAE model)
If the answer is No, move to query 8		• Use ash in conservation agriculture
8. Can it be used for biochar making?	Yes	• Leave stubbles in field • Install biochar klin (e.g. IARI model) • Use biochar in conservation agriculture

Table 1.
Model plan for managing crop residues at local and regional scales.

The Indian Ministry of Agriculture recently created a National Policy for Management of Crop Residue (NPMCR) [7] in addition to the aforementioned. The NPMCR's primary goals are listed below [6]:

1. To reduce the loss of important soil nutrients and to increase the variety of uses for crop residue in industrial applications, promote technology for the best usage and in-situ management of crop residue.

2. Create and encourage the use of suitable crop machinery in agricultural techniques, such as the modification of grain recovery equipment (harvesters with twin cutters to cut the straw). Offer discounts and incentives to encourage the purchase of mechanical sowing equipment such as baling, shredding, and turbo seeders.

3. Work with the National Remote Sensing Agency (NRSA) and the Central Pollution Control Board to monitor agricultural residue management using satellite-based remote sensing technologies (CPCB).

4. Raise money for creative ideas and project proposals using a multidisciplinary approach and fund raising in several ministries.

21. Summary and conclusions

Crop residues are one type of agricultural waste that has presented unique issues because of its enormous volume and lack of management tools. Given that rice and wheat, which typically provide the majority of crop leftover, are the main staples of India, it is apparent that the extensive cultivation of these crops to feed the continuously growing population has resulted in the development of significant amounts of crop residue. India produces 500 Mt. of crop residue annually on average. There is a massive surplus of 140 Mt., out of which 92 Mt. is burned annually, primarily in the northern states such as Punjab, Haryana, and Uttar Pradesh, even though the majority of it is used as fodder, a raw material for energy production, etc. Due to a lack of technical knowledge and appropriate disposal options, small-scale farmers in particular turn to burning crop waste as a cheap alternative. Crop burning on a large scale raises atmospheric CO_2, CO, N_2O, and NOx levels and has caused an alarming rise in air pollution. The air quality in northern India terrifyingly deteriorated, reaching nearly double the allowable Indian threshold and 10 times the WHO standard.

The Indian government has launched numerous programs in an effort to address the issue of crop residue burning. The Indian Ministry of New and Renewable Energy (MNRE) and the Indian Agricultural Research Institute (IARI) are constantly encouraging research and cutting-edge techniques to handle crop waste without burning. Recently developed by the Central Government, the National policy for management of crop residue (NPMCR) outlines laws and regulations that local agencies must follow to address crop burning and promote sustainable management practices. Continued air pollution, particularly in November and December, suggests that the aforementioned restrictions have not effectively stopped crop burning. The true causes of the burning of crop residue are more socioeconomic in nature than agricultural or waste management related. In its place, sustainable alternatives that entail techniques to feed the nutrients in the crop residue back into the same crop areas have been developed. Uncommonly used bio-based products for agricultural waste usage include biogas, charcoal, and in-situ management with mechanical intensification. Composting rules could be developed for rural regions and applied to all farms by farmers associations. Crop residue can be significantly reduced by mechanizing the harvesting process, and farmers may receive equipment subsidies from local authorities. This gap should be filled by the local government, municipality, or farmers' organization, which should also start local assistance programs such as equipment rentals, waste transportation, and possibly linking waste to areas where it can be used as raw materials. Educating the farming community and providing them with socioeconomic and technical support. They must to be informed of the benefits of lower agrochemical costs owing to the use of compost and the additional income they can get from other types of recovery initiatives, such energy production.

Author details

Kawaljeet Kaur* and Preetpal Singh
Department of Agriculture, School of Agriculture, CT Group of Institution Shahpur Campus, Jalandhar, India

*Address all correspondence to: kwlpreet25@gmail.com

References

[1] Aladjadjiyan A. Lessons from Denmark and Austria on the Energy Valorization of Biomass (Contract No: JOU2-CT92-0212, Coordinator for Bulgaria). European Commission: Brussels, Belgium; 1992

[2] Amonette J, Joseph S. Characteristics of biochar: Micro-chemical properties. In: Lehmann J, Joseph S, editors. Biochar for Environmental Management: Science and Technology. Earth Scan: London, UK; 2009. pp. 33-52

[3] Beck-Friis B, Pell M, Sonesson U, Jonsson H, Kirchmann H. Formation and emission of N2O and CH4, from compost heaps of organic household waste. Environmental Monitoring and Assessment. 2000;**62**:317

[4] Franchi E, Agazzi G, Rolli E, Borin S, Marasco R, Chiaberge S, et al. Exploiting hydrocarbon-degrader indigenous bacteria for bioremediation and phytoremediation of amulti-contaminated soil. Chemical Engineering and Technology. 2016;**39**:1676-1684

[5] Gadde B, Bonnet S, Menke C, Garivait S. Air pollutant emissions from rice straw open field burning in India, Thailand and the Philippines. Environmental Pollution. 2000;**157**:1554-1558

[6] Garg S. Bioremediation of agricultural, municipal, and industrial wastes. In: Handbook of Research on Inventive Bioremediation Techniques. 2017

[7] Hatch LE, Luo W, Pankow JF, Yokelson RJ, Stockwell CE, Barsanti KC. Identification and quantification of gaseous organic compounds emitted from biomass burning using two-dimensional gas chromatography-time-of-flight mass spectrometry. Atmospheric Chemistry and Physics. 2015;**15**:1865-1899

[8] Hoornweg D, Bhada-Tata P. What a Waste: A Global Review of Solid Waste Management. Washington, DC, USA: World Bank; 2012

[9] Izaurralde RC, Rosenberg NJ, Lal R. Mitigation of climate change by soil carbon sequestration: Issues ofscience, monitoring, and degraded lands. Advances in Agronomy. 2001;**70**:1-75

[10] Jiang H, Frie AL, Lavi A, Chen JY, Zhang H, Bahreini R, et al. Brown carbon formation from nighttime chemistry of unsaturated heterocyclic volatile organic compounds. Environmental Science & Technology Letters. 2019;**6**(3):184-190.s

[11] Jitendra and Others. India's burning issues of crop burning takes a new turn, down to earth. 2017. Available online: https://www.downtoearth.org.in/coverage/river-of-fire-57924. [Accessed on 7 September 2018]

[12] Lei Z, Chen J, Zhang Z, Sugiura N. Methane production from rice straw with acclimated an aerobicsludge: Effect of phosphate supplementation. Journal of Bioresource Technology. 2010;**101**:4343-4348

[13] Jain N, Bhatia A, Pathak H. Emission of air pollutants from crop residue burning in India. Aerosol and Air Quality Research. 2014;**14**:422-430

[14] Masek OB. Production technologies. 2009. Available online: http://www.geos.ed.ac.uk/sccs/biochar/documents/BiocharLaunch-OMasek.pdf. [Accessed on 6 March 2019]

[15] McHenry MP. Agricultural biochar production, renewable energy generation

and farm carbon sequestration in Western Australia, certainty, uncertainty and risk. Agriculture, Ecosystems & Environment. 2009;**129**:1-7

[16] Lohan SK, Jat HS, Yadav AK, Sidhu HS, Jat ML, Choudhary M, et al. Burning issues of paddy residue management in north-west states of India. Renewable and Sustainable Energy Reviews. 2018;**81**:693-706

[17] Marjanovic I. The best practices for using plant residues, Agrivi. 2016. Available online: http://blog.agrivi.com/post/the-best-practices-for-using-plant-residues. [Accessed on 15 November 2018]

[18] Misra RV, Roy RN, Hiraoka H. On Farm Composting Methods. Food and Agricultural Organization of the United Nations: Rome, Italy; 2003

[19] Mittal SK, Susheel K, Singh N, Agarwal R, Awasthi A, Gupta PK. Ambient air quality during wheat and rice crop stubble burning episodes in Patiala. Atmospheric Environment. 2009;**43**:238-244

[20] Nagendran R. Agricultural waste and pollution. In: Waste. 2011. pp. 341-355

[21] OECD (Organisation for Economic Co-operation and Development). 2001. Available online: https://stats.oecd.org/glossary/detail.asp?ID=77. [Accessed on 10 November 2018]

[22] Perez J, Dorado JM, Rubia TD, Martinez J. Biodegradation and biological treatment of cellulose, hemicellulose and lignin: An overview. Journal of International Microbiology. 2002;**5**:53-56

[23] Pratap Singh D, Prabha R. Bioconversion of agricultural wastes into high value biocompost: A route to livelihood generation for farmers. Adv. Recycl. Waste Manag. 2017:137

[24] Shaxson F, Kassam AH, Friedrich T, Boddey B, Adekunle A. Underpinning conservation agriculture's benefits: The roots of soil health and function. Main Background Document for the Workshop on Investing in Sustainable Crop Intensification: The Case for Improving Soil Health, 22-24 July, 2008. FAO, Rome. 2008

[25] Shilev S, Naydenov M, Vancheva V, Aladjadjiyan A. composting of food and agricultural wastes. In: Oreopoulou V, Russ W, editors. Utilization of By-Products and Treatment of Waste in the Food Industry. New York, NY, USA: Springer; 2006. pp. 283-301

[26] Sequi P. The role of composting in sustainable agriculture. In: The Science of Composting. Dordrecht: Springer; 1996. pp. 23-29

[27] Singh CP, Panigrahy S. Characterization of residue burning from agricultural system in India using a-based observations. Journal of the Indian Society of Remote Sensing. 2011;**39**:423-429

[28] Sommer SG, Dahl P. Nutrient and carbon balance during the composting of deep litter. Journal of Agricultural Engineering Research. 1999;**74**:145

[29] The Hindu Crop Residue-Coal Mix to Nix Stubble Burning. 2018. Available online: http//www.thehindu.com/news/national/other-states/ntpc-to-mix-crop-residue-with-coal-to-curb-crop-burning/article20492123.ece [Accessed on 25 June 2018]

[30] United Nations. Glossary of Environment Statistics, Studies in Methods; Series F, 67. Vol. 96. New York, NY, USA: Department for Economic and Social Information and Policy Analysis, Statistics Division; 1997

[31] Srinivasarao CH, Venkateswarlu B, Lal R, Singh AK, Sumanta K. Sustainable management of soils of dryland ecosystems for enhancing agronomic productivity and sequestering carbon. Advances in Agronomy. 2013;**121**:253-329

[32] Taherzadeh MJ. Ethanol from lignocellulose: Physiological effects of inhibitors and fermentation strategies. [Ph.D. Thesis], Biotechnology, Chemical Reaction Engineering, Chalmers University of Technology, Gothenburg, Sweden. 1999

[33] Wu L, Ma LQ, Martinez GA. Comparison of methods for evaluating stability and maturity of biosolids compost. Journal of Environmental Quality. 2000;**29**:424

[34] https://www.researchgate.net/publication/256378461_Crop_Residues_Management_with_Conservation_Agriculture_Potential_Constraints_and_Policy_Needs

[35] Obi FO, Ugwuishiwu BO, Nwakaire JN. Agricultural waste concept, generation, utilization and management. Nigerian Journal of Technology. 2016;**35**:957-964

[36] NPMCR. Available online: http://agricoop.nic.in/sites/default/files/NPMCR_1.pdf. [Accessed on 6 March 2019]

Chapter 7

Recovery and Characterization of Astaxanthin Complex from Prawn Waste Extract and Its Separation Using Antisolvent Precipitation Technique

Renuka Vinothkumar, Frank Lucien and Janet Paterson

Abstract

Most of the prawns are processed as frozen or cooked prawn meat. The remaining waste (heads, tails and shells) is used as a feed supplement or is directly discarded onto the land by food industries, seafood markets and capture fisheries. Disposal onto the land allows prawn waste to decompose easily in the open air and causes high environmental pollution. At the same time, many valuable compounds present in the waste are lost. It has been accepted that those from marine/brackish waters are considered 'shrimps' while their counterparts from fresh waters are considered 'prawns'. There is a need to generate value-added products from these waste materials from an environmental and economical point of view. The recovery of value-added products from waste material is beneficial in two ways: firstly, to solve the waste disposal problem itself and secondly, to generate additional income. This research particularly focuses on the management of prawn waste and this small-scale research was carried out using the hand-peeled waste of school prawns. The major aim of this research is the recovery and characterization of one of the major valuable components of prawn waste called the 'astaxanthin complex' and its separation from the organic solvent using the antisolvent precipitation technique, which is an innovative approach.

Keywords: prawn waste management, the astaxanthin complex, supercritical CO_2, antisolvent precipitation, volumetric expansion

1. Introduction

The major carotenoid present in prawn waste is the astaxanthin complex. A familiar example is that grey prawn becomes bright orange-red on cooking because the astaxanthin complex including astacene is liberated when cooking denatures the natural protein-pigment complex [1]. The total concentration of the astaxanthin complex in crustaceans including prawns varies with species and location [2]. Free astaxanthin (3, 3′-dihydroxy β, β'-carotene 4, 4′-dione) has two hydroxyl groups,

IntechOpen

one on each terminal ring. The hydroxyl group can react with fatty acid and form esterified astaxanthin. Astaxanthin monoester has one fatty acid attached to one of the hydroxyl groups and astaxanthin diester has one fatty acid attached to each of the hydroxyl groups. The monoesters and diesters are relatively non-polar, whereas free astaxanthin is relatively polar [3]. Thus, the polarity of the astaxanthin complex plays an important role in choosing the optimal solvent for its recovery. Natural sources such as krill, algae and prawn supply astaxanthin as a mixture of free and esterified forms [4, 5]. These extra-functional groups give astaxanthin an extraordinary antioxidant capability and properties unlike other carotenoids [6]. Astaxanthin esters function as powerful antioxidants under both hydrophobic and hydrophilic conditions during experimental *in vitro* studies [7]. However, the astaxanthin complex is sensitive to photo, thermal and oxidative degradation because of the presence of long chain-conjugated double bonds [8].

The astaxanthin complex has attracted considerable interest in recent years because of its powerful antioxidant activity [9]. Mostly the antioxidant properties of the astaxanthin complex are mainly focused on human health benefits [10, 11]. In food and aquaculture, the astaxanthin complex is mainly used as a colour enhancer [6]. In the United States, the astaxanthin complex is a permitted colour additive by the Food and Drug Administration in salmon feed to improve the colour of salmon during farming practices [12]. However, the antioxidant properties of the astaxanthin complex have not been paid much attention in relation to agricultural purposes. In aquaculture feed, the astaxanthin complex extracted from natural materials is preferable [1, 13]. However, the cost of the astaxanthin complex for its use in aquaculture feed is a major concern. Commercially produced synthetic astaxanthin costs about $1000/kg [14]. The high cost and consumer resistance to synthetic pigments have led to the exploitation of natural sources for obtaining the astaxanthin complex. Only a few species of microorganisms (eg: *Haematococcus lacustris*; *Phaffia rhodozyma* and *Chlorella vulgaris*) produce the astaxanthin complex in nature. The astaxanthin complex is also abundant in prawn waste [15]. Therefore, it may be worth recovering the astaxanthin complex from prawn waste and exploring its use in various applications.

Conventional recovery of the astaxanthin complex from natural sources is mainly carried out using solvents. However, there are two major problems associated with the solvent extraction of the astaxanthin complex. The first problem is that solvents also extract other compounds such as lipids present in the natural materials along with the astaxanthin complex. The second problem is that solvent extraction degrades this sensitive pigment [16]. To overcome the problems associated with solvent extraction, an effective way of recovering the astaxanthin complex from natural sources is necessary.

This research dealt with the extraction, characterization and recovery of the prawn pigment, the astaxanthin complex from prawn waste. The astaxanthin complex was extracted from prawn waste using two organic solvents, hexane and acetone. The better solvent for the maximum recovery of the astaxanthin complex was chosen. The compatibility of the solvent with the technique called supercritical antisolvent precipitation was also considered. Characterization of the astaxanthin complex was performed using chromatography techniques such as high-performance liquid chromatography (HPLC), and HPLC coupled with mass spectrometry (HPLC-MS). The astaxanthin complex was then precipitated from the organic solvent using environmentally friendly supercritical carbon dioxide (SCO_2) instead of performing post-extraction steps such as purification or evaporation of the solvent. This application was the novelty of this research.

2. Literature review

2.1 Extraction of astaxanthin from prawn waste

The astaxanthin complex is abundant in prawn waste; however, the concentration of the astaxanthin complex in prawn waste varies with species and location [17–24]. Initially, the extraction of the astaxanthin complex from prawn waste was carried out using vegetable oils. The recovered red oil might then directly be incorporated into the commercial aqua feed [25–29]. The concentration of the astaxanthin complex in the pigmented oil could be further increased by reusing the pigmented oil to extract the astaxanthin complex from fresh waste [22, 26, 30].

Prawn waste is a highly perishable material, which deteriorates rapidly that in turn degrades the astaxanthin complex present in it [31]. Acid treatment and preliminary enzymatic digestion with and without antioxidants have been studied to stabilize the astaxanthin complex in prawn waste prior to oil extraction [17, 24, 31–33]. The recovery of the astaxanthin complex from prawn waste with the aid of preliminary enzymatic digestion depends on the prawn species, fermentation conditions and the type of enzyme used for digestion [24].

The use of organic solvents for the recovery of the astaxanthin complex from prawn waste is best limited to analytical investigations. The traditional solvent extraction processes described in the literature are poorly designed for the commercial recovery of the astaxanthin complex from prawn waste [16]. This is because either the presence of other compounds in the prawn waste, such as lipids, allows the recovery of only a very dilute material, or the solvents used in the extraction of astaxanthin are inefficient and cause degradation of the pigment. Various studies on the solvent extraction of the astaxanthin complex from prawn waste have been investigated [15, 20, 31, 34–36]. Solvent extraction of the astaxanthin complex from natural sources generates more solvent waste and consists of time-consuming complicated post-extraction steps such as evaporation or concentration and purification. The applied heat during extraction and post-extraction may also degrade the astaxanthin complex. To overcome all these possible difficulties associated with solvent extraction and the post-extraction of the astaxanthin complex, a more effective way of recovering the astaxanthin complex from natural sources is necessary.

2.2 Analysis and characterization of the astaxanthin complex from prawn waste

There are two general methods for analyzing the concentration of the astaxanthin complex in a sample: spectrophotometric analysis and HPLC. In both methods, the astaxanthin complex is extracted from a source into a suitable solvent. In spectrophotometric analysis, the light absorbance of the extraction solvent containing the astaxanthin complex is measured at a wavelength that corresponds to the maximum absorbance for the astaxanthin complex (usually between 470 and 480 nm). The concentration of the astaxanthin complex is then calculated using Eq. (1). The problem with the spectrophotometric assay method is that in addition to astaxanthin, other carotenoids or the degradation products of astaxanthin such as astacene will be falsely included as astaxanthin in the results [37–39].

$$[A] = \frac{Xy}{E_{1cm}^{1\%}} \tag{1}$$

where, [A]—concentration of the astaxanthin complex (g); *X*—absorbance at a specific wavelength; *y*—amount of solvent used (mL) and $E_{1cm}^{1\%}$ —extinction coefficient of pure astaxanthin in the extraction solvent.

All those other compounds also absorb light at this wavelength, which in turn leads to an increase in the absorbance measurement. Consequently, this results in an overstatement of the astaxanthin complex concentration in a sample. This overstatement is usually minimized by reporting spectrophotometric analysis results as 'the astaxanthin complex' to indicate that the analysis includes other compounds as well [37]. This is the reason, 'the astaxanthin complex' is the term used in this research rather than 'astaxanthin'. The most technically sound and accurate method for determining the astaxanthin complex concentration of a sample is HPLC, which separates the astaxanthin complex into individual esters, astaxanthin degradation products and other related compounds, to provide accurate concentrations of astaxanthin and its esters [37]. The separation of the astaxanthin complex is achieved by using both normal and reverse phases HPLC and HPLC-MS. This work characterizes the astaxanthin complex from prawn waste extract using normal phase HPLC coupled with atmospheric pressure chemical ionization mass spectrometry (APCI-MS)—negative mode.

2.3 Supercritical antisolvent processes

A supercritical fluid is defined as a substance above its critical temperature and critical pressure. The application of supercritical fluid extraction has been investigated in various industries [40–44]. These fluids can also be used as antisolvents to precipitate materials from conventional solvents where the particle of interest is not soluble in supercritical fluids [45–47]. In supercritical antisolvent processes, the supercritical fluid is used as an antisolvent to bring about precipitation of the substrate(s) dissolved initially in a liquid solvent. This research used carbon dioxide as a supercritical fluid. **Figure 1** shows the working method of the supercritical antisolvent precipitation process. SCO_2 and the organic solvent containing the sample are brought into contact by gradually adding compressed SCO_2 to the solution containing the solute inside a pressure vessel (**Figure 1** Step 1).

SCO_2, which is used as an antisolvent for the solute, initially dissolves in the organic solvent. The organic solvent is completely miscible with carbon dioxide. Upon the addition of carbon dioxide to the organic solvent containing the solute, the antisolvent diffusion decreases the solubility of the solute within the organic phase (**Figure 1** Step 2) and eventually the solute precipitates (**Figure 1** Step 3). The mixture

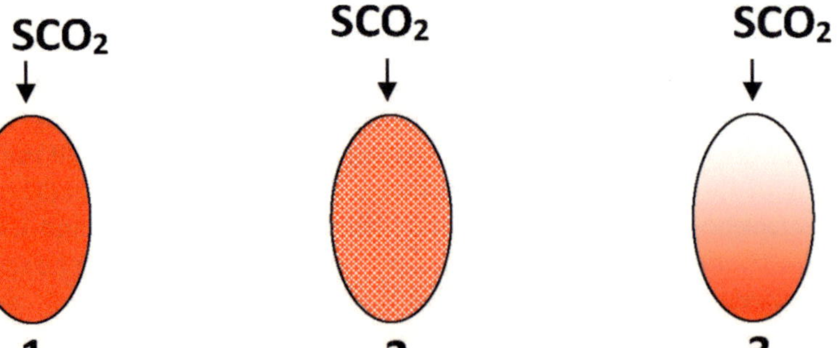

Figure 1.
(Steps 1, 2 and 3) Supercritical antisolvent precipitation technique using SCO_2.

of the solvent and antisolvent can be separated when it is depressurized. Then the liquid solvent and gaseous antisolvent can be recycled [45, 48]. Supercritical antisolvent processes depend on the operating conditions and the affinity of the solute towards the liquid solvent [49]. The success of supercritical antisolvent precipitation relies on the selection of a suitable combination of organic solvent and a supercritical antisolvent for a specific compound of interest [45].

As the aim of this project was the recovery of the astaxanthin complex from prawn waste, the supercritical antisolvent precipitation technique was investigated to separate the pigment from the organic solvent. The use of supercritical fluid as an antisolvent is much simpler than organic liquid antisolvents, which need the application of complex purification processes. There are concerns with the stability of the astaxanthin complex during solvent evaporation. Using the supercritical antisolvent technique, it is possible to precipitate the pigment from the solvent at near ambient temperatures with the proper selection of the antisolvent, thus, avoiding the thermal degradation of the product. And also, supercritical antisolvent processes are carried out in an inert environment and reduced light. This reduces the possible photo- and oxidative degradation of the product [50]. Considering the advantages of the supercritical antisolvent process, we have chosen this technique in the recovery of the astaxanthin complex from prawn waste solvent extract to eliminate the problems associated with the solvent evaporation and purification, and to prevent the degradation of the astaxanthin complex during post-extraction steps.

At the time of this research, SCO_2 was only used as a solvent to extract the astaxanthin complex from crustacean waste [14, 51–53]. The highest (98%) yield of astaxanthin is obtained when the extraction is carried out at low pressure and the highest temperature of 60°C [52]. Therefore, temperature and pressure play an important role in the recovery of astaxanthin from crustacean waste using SCO_2. The SCO_2 extraction process of the astaxanthin complex operates at nearly ambient or low temperatures, eliminating the possibility of heat damage to the pigment and leaving no solvent residue. However, the solubility of carotenoids in SCO_2 is low because of the thick cell wall that resists greater mass transfer and, therefore, the technique may not be effective in extracting total the astaxanthin complex present in the source [14, 51]. The insolubility of the astaxanthin complex in SCO_2 offers the possibility of using SCO_2 as an *antisolvent* in the recovery of the astaxanthin complex from prawn waste. An examination of published studies on supercritical antisolvent precipitation indicates that there have been no previous studies carried out on the recovery of astaxanthin from prawn waste using SCO_2 as an antisolvent. However, some of the studies indicate the application of SCO_2 as an antisolvent in the precipitation of synthetic pigments such as red lake C, pigment yellow 1, pigment blue 15 [54], bronze red pigment [55] and lycopene [50] and grape pomace extract [56] from organic solvents. Therefore, an innovative application was studied in this research to apply SCO_2 as an antisolvent on the recovery of the food and pharmaceutical valuable compound, the astaxanthin complex, present in prawn waste.

The phase behaviour of the supercritical antisolvent with the liquid solvent must be known before antisolvent precipitation. The study of volumetric expansion and vapour–liquid equilibrium data of hexane with SCO_2 at different temperatures in this research offered an idea about the solubility behaviour of SCO_2 in hexane before the supercritical antisolvent precipitation of the astaxanthin complex. The data was also helpful to choose the best operating conditions for the supercritical antisolvent precipitation of the astaxanthin complex.

3. Materials and methods

3.1 Materials used

Due to the difficulty in sourcing adequate commercial prawn waste, raw eastern school prawns (*Metapenaeus macleayi*, approximately 9 cm body length) were obtained from the Department of Primary Industries, Fisheries Conservation Technology Unit, NSW, Australia and hand-peeled to obtain prawn waste to conduct this research (moisture content 74%; Ash 23% and 11% chitin dry basis). This prawn waste was stored at –22°C until used for research. All solvents used were of HPLC grade and all chemicals were of AR grade.

3.2 Recovery of the astaxanthin complex from prawn waste

3.2.1 Solvent extraction of the astaxanthin complex

The astaxanthin complex was extracted from raw as well as freeze-dried prawn waste using hexane and acetone. Raw prawn waste (15 g) was mixed with 30 ml of hexane and acetone in a tightly capped flask separately. The mixture was shaken at 250 rpm under darkness at ambient temperature (23 ± 1°C) for 15 hours to achieve complete extraction of the astaxanthin complex into the solvents. The supernatant, called the pigment extract, was separated from the waste residue and centrifuged at 1100 rpm for 10 minutes. The supernatant was analysed using UV-spectrophotometer at 480 nm to measure the concentration of the astaxanthin complex (See Section 3.2.2).

In the second step, a further 15 g of prawn waste was added to the pigment extract and kept in the shaker under the same experimental conditions for further extraction of the astaxanthin complex. Successive removal of the waste residue from the pigment extract and further addition of prawn waste to the pigment extract was continued to obtain maximum pigment recovery. The concentration of the astaxanthin complex was measured before each addition of prawn waste to the pigment extract. **Figure 1** shows the extraction of the astaxanthin complex from raw prawn waste.

Freeze-dried prawn waste (2.5 g) was mixed separately with 10 ml of hexane and acetone in a tightly capped flask. The pigment extraction was carried out as mentioned above. In the second step, a further 0.5 g of prawn waste was added to the pigment extract (4 ml) and kept in the shaker under the same experimental conditions for further extraction of the pigment. After 20 hours, the concentration of the astaxanthin complex in the pigment extract was measured using UV-spectrophotometer.

3.2.2 Analysis of the astaxanthin complex using UV-spectrophotometer

The synthetic astaxanthin obtained from Sigma-Aldrich is termed 'standard astaxanthin' in this work. UV-spectrophotometer analysis was carried out to measure the concentration of the astaxanthin complex in the pigment extract. The absorbance (A_{480}) of the pigment extract recovered from prawn waste was measured at 480 nm. The concentration of the astaxanthin complex in the pigment extract was calculated using Eq. (1). The calculated specific extinction coefficient of astaxanthin in acetone was 1670. This value was used to calculate the concentration of the astaxanthin complex in the prawn pigment extract. However, the measured specific extinction coefficient of astaxanthin in acetone at a wavelength of 480 nm was different from the one published by [38], where the specific extinction coefficient of astaxanthin in acetone was found to be 2500 at the same wavelength. The specific extinction coefficient of astaxanthin in hexane was not measured because of the insolubility of standard astaxanthin in hexane and also because of the unavailability of standard astaxanthin esters. Therefore, the specific coefficient of astaxanthin in hexane used to calculate the concentration of the astaxanthin complex in the prawn pigment extract was 2100 at a wavelength of 480 nm as mentioned in [57]. The cholesterol and other steroids present in the pigment extract was removed by silica gel column chromatography before the HPLC analysis [8].

3.2.3 Characterization of the astaxanthin complex

3.2.3.1 HPLC analysis of prawn pigment extract

The astaxanthin complex in hexane after the removal of cholesterol and other steroids was analysed by normal phase HPLC to characterize the astaxanthin complex. This experiment was conducted using the same waters HPLC system equipped with a 996-photo diode array detector, which was used to measure the cholesterol content of the pigment extract. HPLC analysis was performed using a Lichrosorb silica gel column (150 mm × 2.1 mm, 5 μm) specially designed for mass spectrometry analytical purposes. Data analysis was performed using Millenium Chromatography Manager software.

The astaxanthin complex in hexane after the removal of cholesterol and other steroids was filtered using a 0.45 μm syringe filter before HPLC analysis. The sample injection volume was 20 μl. A mobile phase of different compositions of hexane: acetone (64:36; 65:35; 84:16; 88:12; 98:2; 99:1, *v*/*v*) was studied with a flow rate 0.5 ml/minute. Absorption spectra were taken in the range of 250–700 nm. Peaks were monitored by UV detection at a wavelength of 480 nm. The percentage of each discrete peak was calculated from the obtained peak area. The mobile phase that gave the better resolution of the astaxanthin complex peaks was chosen and the same mobile phase conditions were used for mass spectrometric analysis.

3.2.3.2 HPLC-APCI-MS analysis

Mass spectrometric study of the astaxanthin complex was performed by HPLC (ThermoFinnigan Surveyor) coupled with an ion trap mass spectrometer (ThermoFinigan LCQ Deca XP Plus) and atmospheric pressure chemical ionization (APCI) source (negative-ion mode). The data analysis was controlled using ThermoFinnigan Xcalibur software. The HPLC conditions and column were the same as the HPLC analysis.

The eluate from the HPLC column was delivered directly to the APCI source, which was set up with the following conditions: vaporizer temperature of 450°C; sheath gas flow rate of 80 ThermoFinnigan arbitrary units; 10 μA discharge current; the capillary temperature of 250°C; 15 V capillary voltage and 30 V tube lens offset. Spectra were acquired over the *m*/*z* range of 550–1300 Da. The analysis was controlled using ThermoFinnigan Xcalibur software.

Astaxanthin monoester and diesters standards were not available. Therefore, different possibilities of negatively charged astaxanthin mono $[(A^{\bullet}–FA)^{-}]$ and diester $[(A^{\bullet}–FA–FA)^{-}]$ ions were calculated manually from the obtained fatty acid profile. Mass spectra of the ester-derived ions were visualized by averaging the data acquired over individual peaks observed in the HPLC chromatograms. The observed *m*/z values were compared to the theoretical ones, facilitating the identification of the astaxanthin species present.

3.2.4 Vapour–liquid equilibrium study of hexane with CO_2

The equilibrium cell consisted of a high-pressure sight gauge with an internal volume of approximately 70 mL. A syringe pump was used for the addition of high-pressure CO_2 to the equilibrium cell. The temperature in the cell was monitored with a type K thermocouple. The equilibrium cell was immersed in a water bath for which the required temperature was maintained using a heater. The system pressure was measured with a pressure transducer with an uncertainty of ±0.035 MPa. The liquid phase was withdrawn from the bottom of the equilibrium cell and recirculated to the top using a metering pump. The solvent trap consisted of a 50 cm^3 sample cylinder. The outlet of the solvent trap was connected to an inverted 5 ml burette (0.1 ml graduations) to determine the quantity of CO_2 present in the liquid and vapour phase samples (**Figure 2**).

Vapour–liquid equilibrium experiments were performed at different temperatures (25°C, 35°C, 45°C and 55°C). During each experiment, the temperature was kept constant by maintaining a constant water bath temperature. Hexane was filled in the sight gauge to a required volume through V6. The system was kept under atmospheric pressure by opening V3 while loading the sample into the equilibrium cell. After sample loading, V3 was closed. The vessel and the connecting lines were purged with low pressure (approximately 5 bar) CO_2 by subsequent opening and closing of V2 to displace any air present. After purging, the equilibrium cell was filled with CO_2 to the desired pressure (approximately 2 bar) by V2 to operate the metering pump and then isolated by closing valves V2 and V3. The metering pump was kept in circulating mode by turning V6 and V4 in their suitable direction. The metering pump was turned on. The system was left for at least 30 minutes to get stable conditions of pressure and

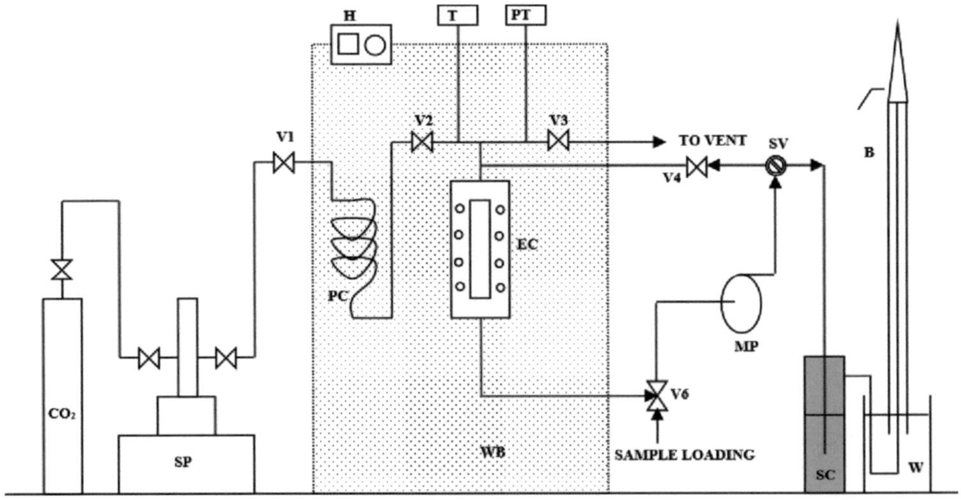

Figure 2.
Schematic diagram of the supercritical antisolvent precipitation setup. (B) burette, (EC) equilibrium cell, (H) heater, (SC) solvent trap (MP) metering pump, (PC) preheating coil, (PT) pressure transducer, (T) thermocouple, (V) valve, (WB) water bath, (CO_2) carbon dioxide cylinder, (SP) syringe pump, (SV) switching valve, (W) container filled with water.

temperature. Then the initial pressure, the water bath and room temperature and the initial level of hexane in the sight gauge were noted. After that CO_2 was added to the sight gauge at several stages by opening and closing V2.

During each stage, samples of the liquid phase were removed from the equilibrium cell to the solvent trap through a switching valve. The CO_2 dissolved in a sample was separated from the hydrocarbon component using a solvent trap filled with a known quantity of isopropanol. The isopropanol in the solvent trap was saturated with CO_2 before sampling. The gas evolved from the liquid phase and was measured in terms of displacement volume with an inverted graduated burette. The liquid component remaining in the connecting line between valve V6 and the solvent trap was recovered by rinsing the connecting line with isopropanol.

The rinse solution was combined with the solvent trap solution and transferred to a flask containing a known mass of internal standard (0.1 g of cumene). The combined solution was then analysed by gas chromatography to determine the total mass of hexane associated with CO_2 from the sample. The use of an internal standard, in this case, cumene, eliminates the need to know the final volume of the combined solution when calculating the total mass of hexane. The mass of hexane in the collected sample was calculated using Eq. (2). Number of moles of hexane in the sample was calculated by dividing the mass of hexane by the molecular weight of hexane.

$$M_H = \frac{1}{RF}\left(\frac{A_H}{A_C} \times M_C\right) \tag{2}$$

where, M_H—mass of hexane; RF—response factor of hexane; A_H—area of hexane; A_C—area of cumene and M_C—mass of cumene.

Since the solvent trap was operated at near atmospheric pressure, the actual quantity of CO_2 collected from the sample was calculated using the ideal gas equation Eq. (3).

$$nc = \frac{\left(P_f V_f - P_i V_i\right)}{RT} \tag{3}$$

where, n_c—number of moles of CO_2; P and V—pressure and gas volume in the solvent trap, respectively; R—universal gas constant and T—temperature. The subscripts '*i*' and '*f*' refer to the initial and final conditions in the solvent trap.

As can be seen from **Figure 2**, the gas space above the hexane in the solvent trap was directly connected to the gas space at the top of the burette. The pressure in the solvent trap was, therefore, deduced from the height of water in the burette. It is important to note that each gas volume term in Eq. (3) was the combined volume of gas in the solvent trap and the burette, corrected for the amount of hexane added initially to the solvent trap. In the case of V_f, a correction was also made for the amount of cumene collected in the solvent trap as well as the volume displacement of water in the burette. A small loss of hexane occurs during the initial saturation of the hexane in the solvent trap with CO_2.

From the number of moles of hexane and CO_2, the mole fraction of hexane and CO_2 in percentage in the collected sample was calculated. The ratio of mole percentage of hexane to mole percentage of CO_2 in the liquid phase was then determined. The composition of the liquid phase was determined from the average of at least three measurements, with a relative standard deviation of less than 5%. The relative standard deviation was calculated with respect to the ratio of mole percentage of hexane to mole percentage of CO_2 in the sample. The drop in pressure in the equilibrium cell was less than 0.1 MPa during the sampling procedure. After each sampling, the desired pressure was restored in the equilibrium cell followed by recirculation of the liquid phase for at least 30 minutes.

3.2.5 Volumetric expansion of hexane using CO_2

The measurement of the volumetric expansion of hexane using CO_2 was carried out using the same apparatus shown in **Figure 1** with some minor modifications. The vapour phase sampling line with the burette was removed from the switching valve. A scale, with 1 mm graduations, was fitted along the visible length of the sight gauge. The apparatus was then calibrated to determine the volume of liquid as a function of the scale length. Volumetric expansion runs were performed at different temperatures (25°C, 35°C, 40°C, 45°C and 55°C). The volumetric expansion data obtained at 25°C and 40°C was used to compare the results with the published data. The temperature was kept constant by maintaining a constant water bath temperature during each run.

Hexane was filled in the sight gauge to a required volume through V6. The system was stabilized for 30 minutes. After stabilization, the initial pressure, the water bath and room temperature and the initial level of hexane in the sight gauge were noted. Then CO_2 was added to the sight gauge at several stages by opening and closing V2. The liquid phase was re-circulated and let come to equilibrium over approximately 30 minutes. For each stage of addition of CO_2, the temperature, pressure and liquid level rose in the sight gauge were noted.

From the initial level of hexane and the level of hexane after the addition of CO_2 in the equilibrium cell, the volumetric expansion of hexane in the equilibrium cell was calculated. The volumetric expansion of hexane at each stage of addition of CO_2 at a given temperature and pressure was calculated using Eq. (4).

$$Expansion(\%) = \frac{V_F - V_I}{V_I} \times 100 \qquad (4)$$

where, V_I—initial volume of hexane in the equilibrium cell and V_F—final volume of hexane after that addition of CO_2

3.2.6 Supercritical antisolvent precipitation of the astaxanthin complex

Supercritical antisolvent precipitation of the astaxanthin complex was performed with the same experimental setup that was used for volumetric expansion experiments. Because the astaxanthin complex was sensitive to heat and light, the experiment was performed under reduced light and the lowest temperature possible (35°C). The experimental pressure and temperature were above the critical temperature and pressure of CO_2. Therefore, CO_2 used in this study was mentioned as SCO_2. The solution containing the astaxanthin complex in hexane was made-up to a concentration of approximately 30 mg/100 mL after the removal of cholesterol and other steroids. This sample solution was filled in the equilibrium cell to the bottom of the graduated scale through V6. The volumetric expansion of hexane containing the astaxanthin complex was calculated at different pressure upon the addition of SCO_2 to the equilibrium cell. This experiment was performed to evaluate the effect of the presence of the astaxanthin complex in hexane on the volumetric expansion using SCO_2. The visual observations were recorded during the experiment.

The profile of the astaxanthin complex precipitated at different pressure during supercritical antisolvent precipitation was studied. The astaxanthin complex in hexane was filled in the equilibrium cell to the bottom of the graduated scale through V6. The pressure inside the cell was increased by adding SCO_2 through V2 (**Figure 2**). The supernatant was collected at different pressure (5.0 MPa, 5.5 MPa, 6.0 MPa and 6.4 MPa) via V4 and analysed using HPLC. From the peak area of the HLPC chromatogram, the composition of the astaxanthin complex in the supernatant at respective pressure was calculated.

The yield of the astaxanthin complex during supercritical antisolvent precipitation was calculated. The astaxanthin complex in hexane was filled in the equilibrium cell to the bottom of the graduated scale through V6. The pressure inside the cell was increased by adding SCO_2 through V2. The supernatant (200 μl) was collected at a pressure of 6.6 MPa into a sample holder containing a known volume of hexane through the switching valve. The concentration of the astaxanthin complex in the collected supernatant was analysed using UV-spectrophotometer. Then the concentration of the astaxanthin complex in the supernatant was calculated and corrected to a suitable dilution factor.

4. Results and discussion

4.1 Solvent extraction of the astaxanthin complex from prawn waste

Figure 3 shows the extraction of the astaxanthin complex from raw as well as freeze-dried prawn waste using hexane and acetone. The concentration of the astaxanthin complex in the pigment extract increased with the extraction time when raw prawn waste was used in hexane. After 54 hours of extraction of raw prawn waste in hexane, the concentration of the astaxanthin complex increased to 4 mg/100 ml in hexane.

When acetone was used for the extraction of the astaxanthin complex from raw prawn waste, the concentration of the astaxanthin complex in the pigment extract was nearly the same as in hexane extraction. However, there was a reduction in the concentration of the astaxanthin complex in the pigment extract after the addition of the

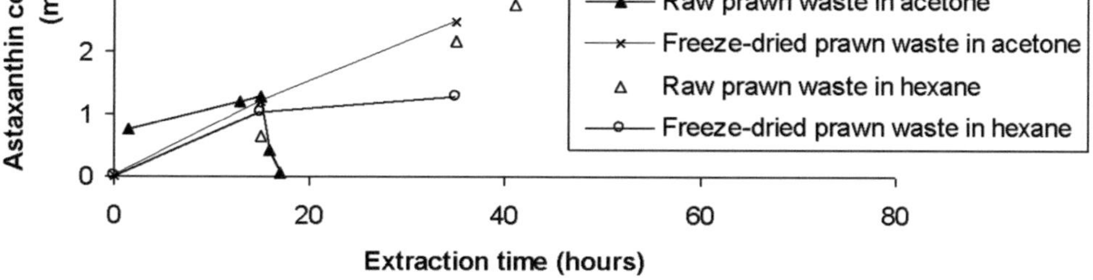

Figure 3.
The astaxanthin complex extracted from raw and freeze-dried prawn waste using hexane and acetone.

second batch of raw prawn waste into the pigment extract. For this reason, hexane was chosen as the better solvent to extract the astaxanthin complex from raw prawn waste.

When freeze-dried prawn waste was used for the recovery of the astaxanthin complex, acetone showed better recovery of the astaxanthin complex than hexane (**Figure 3**). Acetone was the better solvent to extract the astaxanthin complex from freeze-dried prawn waste than hexane; however, this method required freeze-drying of prawn waste.

The concentration of the astaxanthin complex recovered from prawn waste using hexane was nearly equal to the concentration of the astaxanthin complex recovered from freeze-dried prawn waste using acetone. For this research, hexane was chosen as the suitable solvent for the extraction of the astaxanthin complex from raw prawn waste. Using hexane, the astaxanthin complex can be recovered directly from raw prawn waste irrespective of its moisture content, thus eliminating freeze-drying of prawn waste. This solvent extraction method using hexane also provides the possibility of reusing the solvent.

HPLC analysis of the astaxanthin complex dissolved in hexane after silica gel chromatography showed the presence of a free astaxanthin peak [8, 17]. In contrast, experimental trials of this present work indicated that free astaxanthin was not soluble in hexane. Therefore, free astaxanthin peak was not expected during HPLC analysis of the astaxanthin complex dissolved in hexane and the same was confirmed by HPLC-APCI-MS.

4.2 Characterization of the astaxanthin complex by HPLC analysis

There was no separation of peaks when hexane: acetone (64:36 and 65:35, *v*/*v*) was used as a mobile phase. Both the solvent and the astaxanthin complex eluted out together without separation. When the polarity of the mobile phase was reduced slightly (hexane: acetone, 84:16 and 88:12 *v*/*v*), the solvent and the astaxanthin complex started to resolve into discrete peaks although with poor resolution (**Figure 4A** and **B**). When the polarity of the mobile phase was reduced considerably (hexane: acetone, 98:2 *v*/*v*), a good resolution of 12 discrete peaks of the astaxanthin complex was obtained (**Figure 4C**). The percentage composition of each peak was calculated from the peak area. When the composition of hexane in the mobile phase was increased further (hexane: acetone, 99:1, *v*/*v*), the retention time of the peaks was increased (result not shown). Therefore, the mobile phase containing hexane (98%) and acetone (2%) was chosen as the optimum condition for the analysis of the astaxanthin complex.

Most of the diesters (peaks 1–6, 8) were eluted out first followed by monoesters (peaks 3, 9–11). The esters were classified into mono and diesters from the HPLC-APCI-MS results. It should be noted that peak 3 contained both mono- and diesters. The astaxanthin complex recovered from raw prawn waste contained about 71% of monoesters, 10% diesters, 5% impurities and 14% of the unidentified compound when analyzed using normal phase HPLC. In contrast, the pigment extract from prawn waste contained more diesters (76%) and fewer monoesters (18%) and free astaxanthin (6%) when analysed using reverse phase HPLC [17]. In addition, the reverse phase HPLC analysis of the sample dissolved in hexane after silica gel chromatography showed the presence of a free astaxanthin peak [8, 17]. In contrast, experimental trials of this present work indicated that free astaxanthin was not soluble in hexane. Therefore, free astaxanthin peak was not expected during HPLC analysis of the astaxanthin complex in hexane, which was confirmed by HPLC-APCI-MS. Thus, free astaxanthin was not extracted from prawn waste using hexane, and the

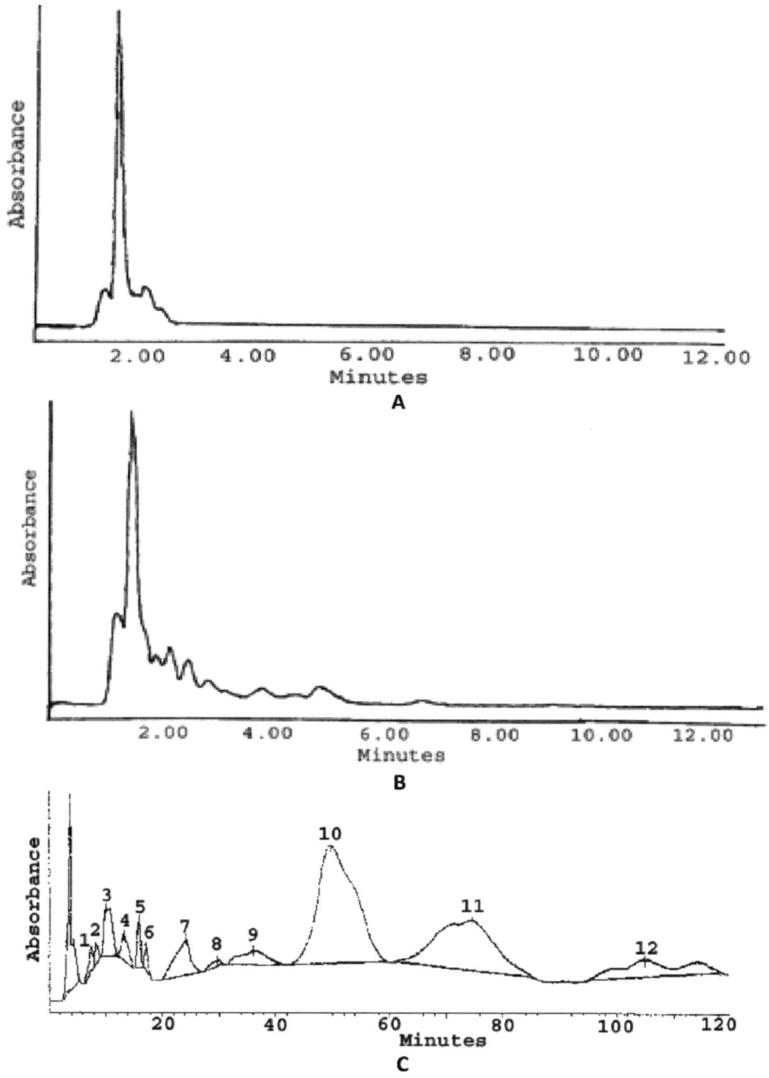

Figure 4.
HPLC analysis of the astaxanthin complex extracted from prawn waste using the mobile phase with different compositions of hexane and acetone (v/v). (A) 84:16 (B) 88:12 (C) 98:2.

astaxanthin complex extracted from prawn waste in hexane was esterified form. The detailed analysis of HPLC-APCI-MS results is not included in this article due to the page number constraints.

4.3 Vapour–liquid equilibrium study of hexane with CO_2

The volumetric expansion of hexane with CO_2 at different temperatures and pressure is shown in **Figure 5**. Each point in this figure represented a single determination of the volumetric expansion of hexane with CO_2 at a given pressure. A smaller volume of liquid in the precipitation chamber allowed a greater degree of expansion. At low pressure, the data were almost a linear function of pressure where the temperature was kept constant. At high pressure, the volumetric expansion increased

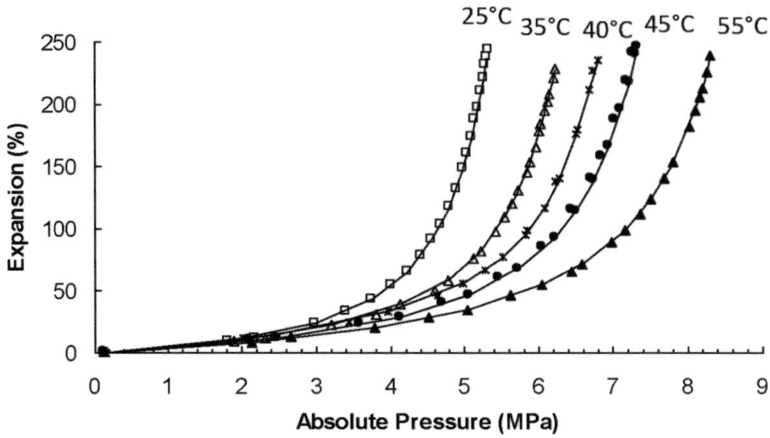

Figure 5.
Expansion of hexane using CO_2 at different temperatures: a binary system.

exponentially as a consequence of the considerable increase in the solubility of CO_2 in this range of pressure at a constant temperature. At constant pressure, hexane showed more expansion at a lower temperature than at a higher temperature. This is attributed to the higher solubility of CO_2 in hexane at lower temperatures [58].

It was chosen that supercritical antisolvent precipitation of the astaxanthin complex was carried out at 35°C. The volumetric expansion of hexane with CO_2 was started around 5 MPa at 35°C (**Figure 5**). Therefore, it was expected that the astaxanthin complex in hexane would precipitate around 5 MPa during antisolvent precipitation using SCO_2. It was necessary to know the solubility data of CO_2 in hexane before supercritical antisolvent precipitation of the astaxanthin complex recovered from raw prawn waste. The increase in the solubility of CO_2 in hexane increases the precipitation of the astaxanthin complex from hexane. This is because the increase in solubility of CO_2 in hexane reduces the affinity of hexane towards the astaxanthin complex, thereby facilitating the precipitation of the astaxanthin complex from hexane.

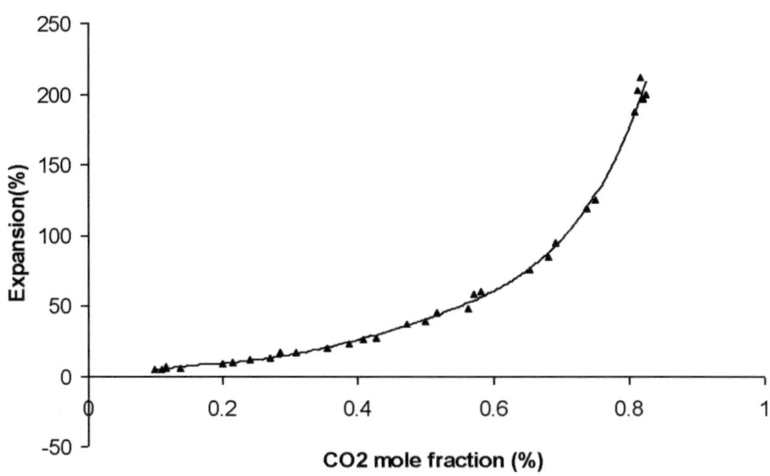

Figure 6.
Relationship between volumetric expansion and the solubility of CO_2 in hexane.

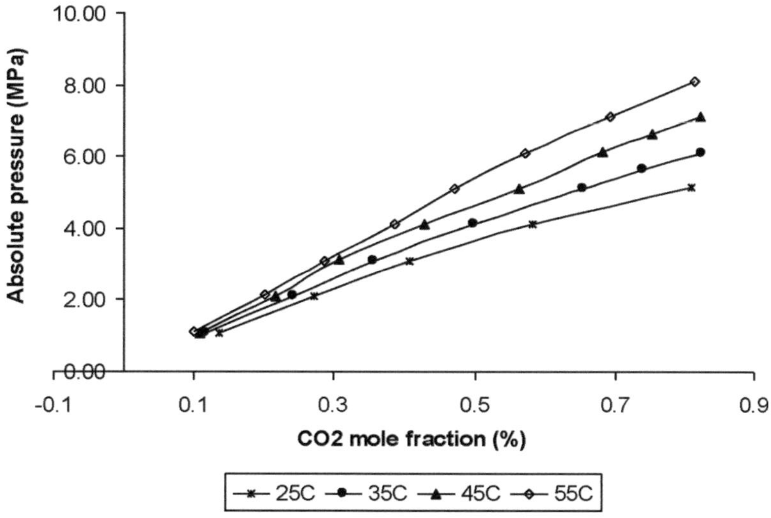

Figure 7.
Solubility of CO_2 mole fraction in hexane at different pressure and temperature.

The relationship between volumetric expansion and the solubility of CO_2 in hexane is shown in **Figure 6**. This indicated that although volumetric expansion was linear when the solubility of CO_2 in hexane was low, the volumetric expansion increased exponentially to the increase in the solubility of CO_2 in hexane. Volumetric expansion curves at different temperatures coincided in a single line when plotted as a function of the solubility of CO_2 in hexane. The volumetric expansion of hexane with CO_2 was initiated around 5 MPa when measured at 35°C (**Figure 5**). The solubility of CO_2 at that pressure was about 0.6 (**Figure 7**), where the precipitation of the astaxanthin complex was started during supercritical antisolvent precipitation. An increase in pressure increased the solubility of CO_2 in hexane, which in turn increased the volumetric expansion of hexane exponentially (**Figure 6**).

Therefore, the rate of precipitation of the astaxanthin complex increased with an increase in pressure during supercritical antisolvent precipitation. At a pressure of 6.6 MPa, the solubility of CO_2 hexane was over 0.8. Thus, the precipitation yield of 92.5% of the astaxanthin complex was obtained at this pressure range during supercritical antisolvent precipitation.

4.4 Supercritical antisolvent precipitation of the astaxanthin complex

The volumetric expansion of pure hexane and the astaxanthin complex in hexane using SCO_2 was compared at 35°C (**Figure 8**). The volumetric expansion of pure hexane using SCO_2 was a binary system. The astaxanthin complex in hexane was assumed to be a ternary system. In this case, the astaxanthin complex was assumed to be a solid phase as a whole even though the astaxanthin complex contained a mixture of astaxanthin esters, impurities and unidentified compound(s). The volumetric expansion of the ternary system (the astaxanthin complex, hexane and SCO_2) was in close agreement with the volumetric expansion of the binary system (pure hexane and SCO_2).

As expected from the volumetric expansion data of hexane with SCO_2, the precipitation of the astaxanthin complex started at 5 MPa. For a solution, the volumetric expansion of the liquid phase should be independent of the initial volume of the

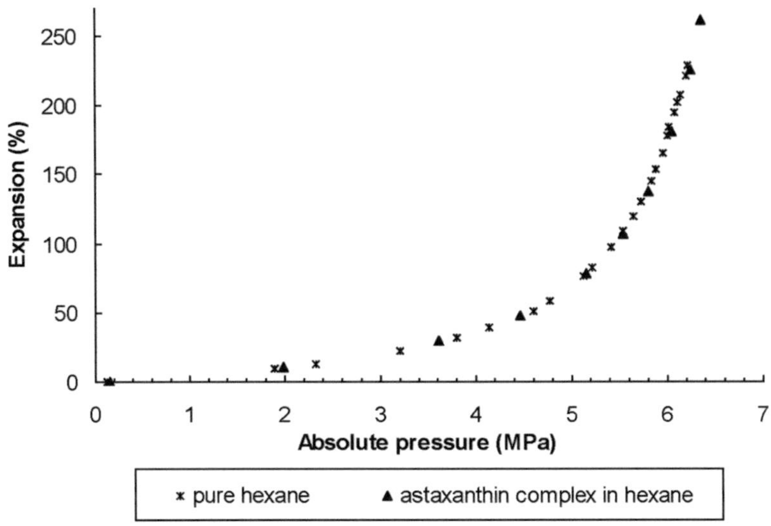

Figure 8.
Comparison of volumetric expansion of pure hexane (a binary system) and hexane containing the astaxanthin complex (a ternary system) using SCO_2 at 35°C.

solution once precipitation has occurred at a given temperature and pressure [59]. The data obtained from this work were consistent with this principle. This study also showed that the presence of the astaxanthin complex in hexane at a concentration of approximately 30 mg/100 mL did not affect the expansion behaviour of hexane with SCO_2 significantly. This type of behaviour has been noted for other ternary systems such as dimethylsulfoxide containing yttrium acetate with SCO_2 [60] while in some cases dimethylsulfoxide containing cefanoid with SCO_2, an increase of the mixture critical pressure due to the presence of the solute has been evidenced [61].

Based on the HPLC-MS ionization analysis, the precipitation profile of the astaxanthin complex at different pressures during the supercritical antisolvent precipitation using SCO_2 is given in **Table 1**. At lower pressure (5 MPa), most of the impurities and part of the unidentified compound(s) were precipitated. Impurities were fully precipitated at 5.5 MPa. The precipitation of astaxanthin monoesters was observed at the intermediate pressure (5.5 MPa). Eventually, diesters were precipitated at higher pressure (6.4 MPa).

The polarity of the astaxanthin complex in hexane played a major role during supercritical antisolvent precipitation using SCO_2. Non-polar material, the astaxanthin complex was dissolved in the non-polar solvent, hexane. As the polarity of

Absolute pressure (MPa)	The precipitate obtained during supercritical antisolvent precipitation
5.0	Impurities and unidentified compound*
5.5	Monoesters*
6.4	Diesters*

**Some proportions of the unidentified compounds remain in the supernatant throughout the precipitation process.*

Table 1.
Summary of the astaxanthin complex precipitation at different pressures during supercritical antisolvent precipitation.

impurities and unidentified compounds in hexane was relatively high compared to astaxanthin esters, these compounds were precipitated first at lower pressure. This is because non-polar hexane has less affinity toward polar compounds than non-polar compounds.

The polarity of astaxanthin monoesters was high compared to astaxanthin diesters because astaxanthin diesters are attached to two fatty acid end groups whereas astaxanthin monoesters have only one fatty acid end group. So, astaxanthin diesters have a stronger affinity towards hexane than astaxanthin monoesters. Monoesters were, therefore, precipitated first from the astaxanthin complex at intermediate pressure followed by astaxanthin diesters at high pressure. Non-polar hexane did not extract the unesterified astaxanthin or free astaxanthin because of its polar nature. The proportion of free astaxanthin present in the prawn waste could not be worked out using hexane as an extraction solvent.

During supercritical antisolvent precipitation using SCO_2, some portion of the unidentified compound(s) remained in the sample even at higher pressure with no precipitation. The reason for this is still not very clear. However, this indicated that the unidentified compound(s) present in the prawn extract could be a mixture of several compounds. The yield study showed that 92.5% of the astaxanthin complex was precipitated at a pressure of 6.6 MPa. Further increase in pressure will increase the yield of the astaxanthin complex precipitation during supercritical antisolvent precipitation using SCO_2. The yield of the astaxanthin complex obtained from published solvent extraction methods cannot be compared with this experiment because of the difference in the extraction method and prawn species used. Thus, an innovative method was developed in this research to recover the astaxanthin complex from prawn waste using SCO_2. As the application is new, the results of this work cannot be compared with the published work.

5. Conclusions

Extraction of the astaxanthin complex from prawn waste using hexane offers an easy way of recovering the astaxanthin complex from prawn waste. The astaxanthin complex can be recovered directly from raw prawn waste irrespective of its moisture content eliminating the need for sample preparation steps such as freeze-drying. As the astaxanthin complex extraction in hexane is carried out at ambient temperature ($23 \pm 1°C$), the energy required for heating is eliminated. This method also offers the possibility of reusing the solvent, hexane.

Characteristic study of the astaxanthin complex offers the separation of the astaxanthin complex into discrete fractions of either monoesters or diesters using normal-phase HPLC with a mobile phase consisting of hexane and acetone. Reliable identification of the type of astaxanthin ester in the astaxanthin complex can be achieved without prior derivatization by directly coupling HPLC with APCI-MS. This analytical method using normal phase HPLC-APCI-MS permits the isolation and identification of previously unreported impurities in the astaxanthin complex extracted from prawn waste.

The precipitation of the astaxanthin complex from hexane using environmental friendly SCO_2 is the novelty in this research. The method is simple, easy and time-effective. This method eliminates the post-extraction steps such as purification, evaporation of the solvent and the associated time. Impurities can be precipitated at lower pressure during supercritical antisolvent precipitation of the astaxanthin complex instead of performing separate purification steps. It offers the recyclability of hexane as well as CO_2. Heating hexane for the extraction process requires enormous amounts of energy that in turn produces greenhouse gas emissions. However, this method did not use a heating source. Therefore, this method saves energy and its associated costs.

This method may also reduce the degradation of the pigment as the precipitation of the astaxanthin complex occurs at low temperatures (35°C), with reduced light and an oxygen-free environment. The astaxanthin complex recovered from prawn waste can be used in aquaculture, food and pharmaceutical applications. Some of the processing steps can be omitted depending on the end use of the recovered astaxanthin complex. At last, this technique is not limited to recovering the astaxanthin complex from prawn waste only. The same technique can be used to recover the astaxanthin complex from other crustacean waste and other major producers of astaxanthin including microalgae.

6. Recommendations

A detailed characteristic study is required to identify the unidentified compound(s) present in the astaxanthin complex. Further improvement in the instrumental setup is required to effectively collect the pigment after precipitation with a view to the possible commercialization of this technique. The actual pressure that separates one form of the astaxanthin complex from others during supercritical antisolvent precipitation can be examined. This will offer an easy way to collect individual species of the astaxanthin complex and will facilitate the study of the stereoisomeric profile of each form of collected astaxanthin complex. These studies are best carried out while optimizing pilot-scale extraction and purification, a scale beyond the scope of this project.

Acknowledgements

I acknowledge Prof. Wilem F. Stevens, Mahidol University, Thailand, who guided me during this research, when required. I also acknowledge Dr Russel Pickford and Dr John Craske from UNSW for their support with the HPLC-MS analytical work.

Notes/thanks/other declarations

Many Thanks to the Department of Primary Industries, Fisheries Conservation Technology Unit, NSW, Australia for supplying raw Eastern School prawns for the research. I also thank the Australian Government for the financial support by providing the International Postgraduate Research Scholarship to undertake this research.

Author details

Renuka Vinothkumar*, Frank Lucien and Janet Paterson
University of New South Wales, Sydney, Australia

*Address all correspondence to: renuevino@yahoo.com

References

[1] Britton G. General carotenoid methods. Methods in Enzymology. 1985;**III**:113-149

[2] Lambertsen G, and Braekkan, O. R. Method of analysis of astaxanthin and its occurrence in some marine products. Journal of the Science of Food and Agriculture. 1971;**22**:99-101

[3] Thomason M. HPLC and spectrophotometric analysis of carotenoids from *Haematococcus* algae oleoresi. In: Cyanotech Corporation, editor. 2001. pp. 1-9

[4] Kitahara T. Behavior of carotenoids in the chum salmon (*Oncorhynchus keta*) during anadromous migration. Comparative Biochemistry and Physiology. 1983;**76B**:97-101

[5] Takaichi S, Matsui K, Nakamura M, Muramatsu M, Hanada S. Fatty acids of astaxanthin esters in krill determined by mild mass spectrometry. Comparative Biochemistry and Physiology. 2003; **Part B 136**:317-322

[6] Lorenz RT. NatuRose TM natural astaxanthin (*Haematococcus* algae meal) as a carotenoid and vitamin source for salmonids. In: Cyanotech Corporation, editor. 2000. pp. 1-24

[7] Kobayashi M, Sakamoto Y. Singlet oxygen quenching ability of astaxanthin esters from the green alga *Haematococcus pluvialis*. Biotechnology Letters. 1999;**21**:265-269

[8] Lin WC, Chien, J.T. and Chen, B.H. Determination of carotenoids in spear shrimp shells (*Parapenaeopsis hardwickii*) by liquid chromatography. Journal of Agriculture and Food Chemistry. 2005;**53**:5144-5149

[9] Palozza P, and Krinsky, N.I. Astaxanthin and canthaxanthin are potent antioxidants in a membrane model. Archives of Biochemistry and Biophysics. 1992;**297**(2):291-295

[10] Terao J. Antioxidant activity of β-carotene-related carotenoids in solution. Lipids. 1989;**24**(7):659-661

[11] Lima SGM, Freire MCLC, Oliveira VdS, Solisio C, Converti A, de Lima ÁAN. Astaxanthin delivery systems for skin application: a review. Marine Drugs. 2021;**19**(9):511

[12] Turujman SA, Wamer WG, Wei RR, Albert RH. Rapid liquid chromatographic method to distinguish wild salmon from aquacultured salmon fed synthetic astaxanthin. Journal of AOAC International. 1997;**80**(3):622-632

[13] Meyers SP, Bligh D. Characterization of astaxanthin pigments from heat-processed crawfish waste. Journal of Agriculture and Food Chemistry. 1981;**29**(3):505-508

[14] Charest DJ, Balaban MO, Marshall MR, Cornell JA. Astaxanthin extraction from crawfish shells by supercritical CO_2 with ethanol as cosolvent. Journal of Aquatic Food Product Technology. 2001;**10**(3):79-91

[15] Armenta LR, Guerrero LI, Huerta S. Astaxanthin extraction from shrimp waste by lactic fermentation and enzymatic hydrolysis of the caroteno-protein complex. Journal of Food Science. 2002;**67**(3):1002-1006

[16] Gildberg A, Stenberg E. A new process for advanced utilization of shrimp waste. Process Biochemistry. 2001;**36**:809-812

[17] Guillou A, Choubert G, Noue J. Separation and determination of carotenoids, retinol, retinal and their dehydro forms by isocratic reverse phase HPLC. Food Chemistry. 1993;**47**:93-99

[18] Okada S, Nur-E-Borhan SA, Yamaguchi K. Carotenoid composition in the exoskeleton of commercial black tiger prawns. Fishery Science. 1994;**60**:213-215

[19] Chakrabarti R. Carotenoprotein from tropical brown shrimp shell waste by enzymatic process. Food Biotechnology. 2002;**16**(1):81-90

[20] Sachindra NM, Bhaskar N, Mahendrakar NS. Recovery of carotenoids from shrimp waste in organic solvents. Waste Management. 2006;**26**:1092-1098

[21] Sachindra NM, Mahendrakar NS. Process optimization for extraction of carotenoids from shrimp waste with vegetable oils. Bioresource Technology. 2005;**96**(10):1195-1200

[22] Sachindra NM, Bhaskar, N., and Mahendrakar, N. S. Carotenoids in different body components of Indian shrimps. Journal of the Science of Food and Agriculture. 2005a;**85**(1):167-172

[23] Holanda HD, Netto FM. Recovery of components from shrimp (*Xiphopenaeus kroyeri*) processing waste by enzymatic hydrolysis. Journal of Food Science. 2006;**71**:298-303

[24] Babu CM, Chakrabarti R, KRS S. Enzymatic isolation of carotenoid-protein complex from shrimp head waste and its use as a source of carotenoids. LWT - Food Science and Technology. 2008;**41**:227-235

[25] Anderson LK, inventor; Bioproducts Incorporated, Warrenton, Oreg., assignee. Extraction of Carotenoid Pigment from Shrimp Processing Waste. USA. 1975. Available from: https://patents.google.com/patent/US3906112A/en

[26] Chen HM, Meyers SP. Extraction of astaxanthin pigment from crawfish waste using a soy oil process. Journal of Food Science. 1982;**47**:892-896

[27] Chen HM, Meyers SP. Ensilage treatment of crawfish waste for improvement of astaxanthin pigment extraction. Journal of Food Science. 1983;**48**:1516-1520

[28] Chen HM, Meyers SP, Hardy RW, Biede SL. Color stability of astaxanthin pigmented rainbow trout under various packaging conditions. Journal of Food Science. 1984;**49**:1337-1340

[29] Shahidi F, Synowiecki J. Isolation and characterization of nutrients and value-added products from snow crab (*Chinoecetes opilio*) and shrimp (*Pandalus borealis*) processing discards. Journal of Agriculture and Food Chemistry. 1991;**39**(8):1527-1532

[30] Handayani AD, Sutrisno NI, Ismadji S. Extraction of astaxanthin from giant tiger (Panaeus monodon) shrimp waste using palm oil: Studies of extraction kinetics and thermodynamic. Bioresource Technology. 2008;**99**(10): 4414-4419

[31] Sachindra NM, Bhaskar N, Siddegowda GS, Sathisha AD, Suresh PV. Recovery of carotenoids from ensilaged shrimp waste. Bioresource Technology. 2007;**98**:1642-1646

[32] Guillou A, Khalil M, Adambounou L. Effects of silage preservation on astaxanthin forms and fatty acid profiles of processed shrimp (*Pandalus borealis*) waste. Aquaculture. 1995;**130**:351-360

[33] Simpson BK, and Haard, N. F. The use of proteolytic enzymes to extract carotenoproteins from shrimp waste.

Journal of Applied Biochemistry. 1985;7:212-222

[34] Mandeville S, Yaylayan V, Simpson BK, Ramaswamy H. Isolation and identification of carotenoid pigments, lipids and flavor active compounds from raw commercial shrimp waste. Food Biotechnology. 1991;**55**:185-196

[35] Healy MG, Green AT, editor. Bioprocessing of marine crustacean (*Nephrops norvegicus*) shell waste. In: Bionet Annual Conference 2003; Galway-Mayo Institute of Technology, Ireland; 2003

[36] López-Cervantes J, Sánchez-Machado DI, Gutiérrez-Coronado MA, Ríos-Vázquez NJ. Quantification of astaxanthin in shrimp waste hydrolysate by HPLC. Biomedical Chromatography. 2006;**20**(10):981-984

[37] Cysewski GR. Analytical Methods for Measuring Astaxanthin: Nutritional Outlook. 04 April 2006. Available from: https://www.nutritionaloutlook.com/view/analytical-methods-measuring-astaxanthin

[38] Davies B, H. Chemistry and biochemistry of plant pigments. In: Goodwin TW, editor. Carotenoids. 2nd ed. London, New York, San Francisco: Academic Press; 1976. pp. 38-165

[39] Pfeifhofer HW, Willfurth R, Zorn M, Kranner I. Analysis of chlorophylls, carotenoids, and tocopherols in lichens. In: Kranner IC, Beckett RP, Varma AK, editors. Protocols in Lichenology: Culturing, Biochemistry, Ecophysiology and Use in Biomonitoring. Berlin, Heidelberg: Springer; 2002. pp. 363-378

[40] Coelho JP, Cristino AF, Matos PG, Rauter AP, Nobre BP, Mendes RL, et al. Extraction of volatile oil from aromatic plants with supercritical carbon dioxide: experiments and modeling. Molecules. 2012;**17**(9):10550-10573

[41] Tutek K, Masek A, Kosmalska A, Cichosz S. Application of fluids in supercritical conditions in the polymer industry. Polymers (Basel). 2021;**13**(5):729

[42] Wang W, Rao L, Wu X, Wang Y, Zhao L, Liao X. Supercritical carbon dioxide applications in food processing. Food Engineering Reviews. 2020;**13**(3):570-591

[43] King JW. Modern supercritical fluid technology for food applications. Annual Review of Food Science and Technology. 2014;**5**(1):215-238

[44] Klettenhammer S, Ferrentino G, Morozova K, Scampicchio M. Novel technologies based on supercritical fluids for the encapsulation of food grade bioactive compounds. Foods. 2020;**9**(10):1395

[45] Yeo SD, Kiran E. Formation of polymer particles with supercritical fluids: a review. Journal of Supercritical Fluids. 2005;**34**:287-308

[46] Viera I, Pérez-Gálvez A, Roca M. Bioaccessibility of marine carotenoids. Marine Drugs. 2018;**16**(10):397

[47] Franco P, De Marco I. Supercritical antisolvent process for pharmaceutical applications: a review. Processes. 2020;**8**(8):938

[48] Fusaro F, Hänchen M, Mazzotti M. Dense gas antisolvent precipitation: a comparative investigation of the GAS and PCA techniques. Industrial & Engineering Chemistry Research. 2005;**44**(5):1502-1509

[49] Reverchon E. Supercritical antisolvent precipitation of micro- and

nano-particles. Journal of Supercritical Fluids. 1999;**15**:1-21

[50] Miguel F, Martin, A., Gamse, T., and Cocero, M. J. Supercritical anti solvent precipiration of lycopene: effect of the operating parameters. Journal of Supercritical Fluids. 2006;**36**:225-235

[51] Felix-Valenzuele L, Higuerea-Ciapara I, Goycoolea-Valencia F, Arguelles-Monal W. Supercritical CO_2/ethanol extraction of astaxanthin from blue crab (*Callinected sapidus*) shell waste. Journal of Food Process Engineering. 2001;**24**(2):101-112

[52] López M, Arce L, Garrido J, Ríos A, Valcárcel M. Selective extraction of astaxanthin from crustaceans by use of supercritical carbon dioxide. Talanta. 2004;**64**(3):726-731

[53] Nobre B, Marcelo F, Passos R, Beirao L, Palavra A, Gouveia L, et al. Supercritical carbon dioxide extractionof astaxanthin and other carotenoids from the microalga *Haematococcus pluvialis*. European Food Research Technology. 2006;**223**:787-790

[54] Gao Y, Mulenda TK, Shi YF, Yuan WK. Fine particles preparation of Red lake C pigment by supercritical fluid. Journal of Supercritical Fluids. 1998;**13**:369-374

[55] Hong L, Guo J, Gao Y, Yuan WK. Precipitation of microparticulate organic pigment powders by a supercritical antisolvent process. Industrial & Engineering Chemistry Research. 2000;**39**:4882-4887

[56] Cardoso FAR, Rezende RVP, Almeida RA, Mezzomo N, Ferreira SRS, Meier HF, et al. CFD-based modeling of precipitation by supercritical anti-solvent process of microparticles from grape pomace extract with population balance approach. The Journal of Supercritical Fluids. 2018;**133**:519-27

[57] Barbosa MJ, Morais R, Choubert G. Effect of carotenoid source and dietary lipid content on blood astaxanthin concentration in rainbow trout (*Oncorhynchus mykiss*). Aquaculture. 1999;**176**:331-341

[58] Phiong HS, Lucien FP. Volumetric expansion and vapour-liquid equlibria of α-methylstyrene and cumene with carbon dioxide at elevated pressure. Journal of Supercritical Fluids. 2003;**25**:99-107

[59] Rajasingam R, Tran, T., and Lucien FP. Precipitation of potassium aurocyanide from dipolar aprotic solvents using a supercritical antisolvent. Journal of Supercritical Fluids. 2005;**33**:69-76

[60] Reverchon E, Caputo G, De Marco I. The role of phase behaviour and atomization in the supercritical antisolvent precipitation. Industrial & Engineering Chemistry Research. 2003;**42**:6406-6414

[61] Reverchon E, and De Marco, I. . Supercritical antisolvent micronization of cefonicid: thermodynamic interpretation of results. Journal of Supercritical Fluids. 2004;**31**:207-15

Chapter 8

Activated Charcoal: A Novel Utility Product for Enhanced Animal Health and Production from Agricultural Wastes (Pig Dung and Palm Oil Wastes)

Stephen Nnaemeka Okey and Cosmas Chikezie Ogbu

Abstract

Feed remains the most important component of the cost of production in any poultry operations. Hence, the need to harness the potentials of numerous alternative ingredients such as activated charcoal (AC) produced from agricultural wastes. The objective of the present study was to evaluate the physic-chemical properties of AC produced from a blend of agro-wastes for use as feed additives. 120 day old broilers were distributed into four treatment groups (T1-T4) of thirty birds each and were maintained on a starter and finisher diet for 3 weeks each respectively with T1 (control) fed diet containing no AC. T2-T4 was fed diet which contained 0.5, 1.0 and 1.5% of AC respectively. Physical and chemical properties of the produced AC were determined while the blood and performance parameters were determined and all data subjected to statistical analysis. The AC significantly ($P > 0.05$) reduced feed intake, jejunum pH, FCR, serum cholesterol levels and increased ($P < 0.05$) live weight gain, intestinal lengths, carcass weight and some hematological indices especially in T3 when compared with broilers fed control diet. It was concluded that AC enhanced production and health by improving on the performance, hematology of young chicks and reduction in serum cholesterol level.

Keywords: agricultural wastes, activated charcoal, palm fruit fibre, palm kernel shell, pig dung

1. Introduction

Feed is the most important component in any intensive poultry operation, representing about 70% of the total cost of production [1]. Hence, there is a compelling need to harness the potential of numerous alternative ingredients such as activated charcoal (AC) produced from agricultural wastes as a replacement for expensive conventional ingredients. Agricultural wastes are defined as residues from the growing and processing of raw agricultural products such as fruits, vegetables, meat, poultry, dairy products, and crops [2]. These are renewable resources whose utilization has received great attention due to environmental considerations and the increasing demand for energy worldwide [3]. Agricultural by-products are also being advocated for the production of adsorbents such as activated charcoal (AC) due to their carbon content and the possibility of mitigating environmental pollution through such a process [4]. They have also been found to be renewable and relatively less expensive when compared to other activated charcoal precursors of industrial and petroleum origin such as wood, coal, and lignite [5]. Recently, interests have focused on the use of agricultural wastes such as corn cob, groundnut shell, poultry litter, rice husk, palm kernel shell (PKS), coconut shell, and many others in the production of value-added products such activated charcoal, fertilizer, batteries, biofuel, bio-oil, and biogas [6].

The expansion of agricultural production has naturally resulted in increased quantities of livestock waste, agricultural crop residues, and agro-industrial by-products. The generation of agricultural waste will continue to increase globally as developed and developing countries continue to intensify their farming systems. Research in 2005 revealed that the biomass potential of Nigeria stood at 13 million hectares of fuel wood, 61 million tonnes per year of animal waste, and 83 million tonnes of crop residues [7]. The annual production of agricultural wastes is this high because about 94% and 68% of household are engaged in crop and livestock farming, respectively [8]. The major agricultural crops biomass feedstocks with sustainable potential in Nigeria are millet, yam, cassava, sorghum, rice, groundnut, oil palm, sugar cane, and soybeans [9]. On livestock, the estimates made in 2001 gave the total number of cattle, sheep, goats, horses, pigs, and poultry in Nigeria as 245 million, which altogether produce 0.78 million tonnes of animal waste daily as reported by Akorede et al. [8].

Oil palm industry has been recognized for its contribution toward economic growth and development in Nigeria and Malaysia. That notwithstanding, it has also contributed to environmental pollution due to the production of large quantities of waste products during the product extraction. During the processing of palm oil, more than 70% (by weight) of the processed fresh fruit bunch was left over as oil palm waste consisting majorly of extracted fiber and palm kernel shell [10]. Palm kernel Shell (PKS) are the shell fractions left over after the nut has been removed after crushing in the oil palm mill. There is a surplus of these by-products in the palm oil value chain but their utilization is extremely very negligible. Apart from a few isolated cases where they serve as a source of fuel in cooking, the PKS, for example, are usually dumped in the open field and water ponds which impact negatively on the environment [10]. Therefore, the production of activated charcoal from palm kernel shells using the process of pyrolysis could be a value addition to palm oil processing which is a veritable economic activity in Nigeria and Malaysia [11].

Pyrolysis of agricultural waste is desirable as a large part of the crop body is non-edible and goes as waste. Straw makes up to 50% of the yield of cereal crops and has more potential for char production compared to wood [12]. Different varieties of agricultural wastes that have been tested for pyrolysis include cotton cocoon, groundnut shell, nutshell, palm kernel shell, corn stalk, bagasse, banana leaves, cotton seedcake, garlic stem, pepper stem, tobacco waste, sunflower bagasse, sorghum bagasse, and cassava peels (**Figure 1**) [12, 13].

AAFCO [14] defined activated charcoal as a dark-colored porous form of carbon made from organic parts of plant or animal substances by their incomplete combustion. They are processed carbon materials that are capable of adsorbing various

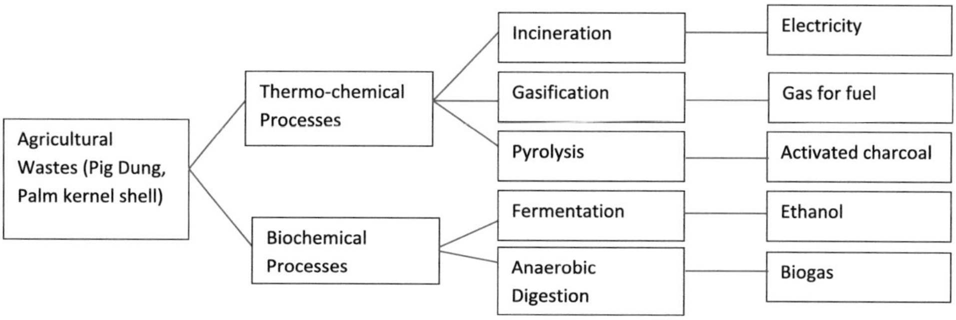

Figure 1.
Methods of conversion of agricultural waste to produce activated charcoal and other products using thermo-chemical and biochemical processes.

substances because of their highly developed pore structure and large internal specific surface area [15, 16]. It is differentiated from elemental carbon by its high surface area and the oxidation of the carbon atoms found at both its outer and inner surfaces [17]. The surface chemistry of activated charcoal confers on it the ability to absorb many gases, aqueous liquid, and poisons [16, 18, 19]. Several studies have shown that activated charcoal is harmless even when it is accidentally consumed, inhaled, or comes in contact with the skin. Although no allergic effects have been associated with its use, the American Academy of Clinical Toxicology, AACT [20] however recommended that activated charcoal should not be taken longer than 12 weeks without stopping. When mixed with water and swallowed to counteract poisoning, activated charcoal adsorbs the poison or drug, inactivating it, and then carries it inert through the entire length of the digestive tract out of the body [16, 19, 21].

Majewska [22] carried out experiment to determine the effect of hardwood charcoal supplementation on the performance and carcass characteristics of broiler at varying inclusion level in the diet. The results showed that at 3% dietary supplementation, the birds were 5% and 3.5% heavier than the control and the dressing percentages and the relative weights of the muscles were also improved at 21 and 42 days, respectively. The author attributed the results to the detoxifying effects of charcoal, thereby lowering the surface tension of the intestinal digest to support liver function with respect to fat digestion. More so, the adsorption properties of charcoal act curatively on the gastrointestinal tract (GIT), adsorbing gases such as hydrogen sulphide and ammonia that are formed there, including bacterial toxins and mycotoxins produced by fungi [23]. Jiya [24] supplemented activated charcoal at 0.5% in broiler feeds and noted increased relative organ weights and reduced cholesterol level of carcass which he attributed to efficient mineral uptake and nutrient utilization.

Durunna et al. [25] reported improved growth rates and reduced flatulence, fly population, and litter odor at varied inclusion levels of wood charcoal in the feed of broiler birds. In another research by Dim et al. [26] to ascertain the effect of dietary supplementation of activated charcoal on growth, hematology, and serum lipid profiles of broilers, the final body weight, average daily weight, and FCR favored birds placed on 6% charcoal inclusion than other groups and the control after 56 days trial period.. More so, Dim et al. [26] noted that the white blood cell (WBC) count and the packed cell volume (PCV) were not affected at both the starter and finisher phases. However, the hemoglobin concentration (Hb) and the red blood cell count were significantly improved [27, 28], while the cholesterol and lipoprotein levels were significantly reduced with no effect on triglyceride at both phases [29].

2. Materials and methods

2.1 Ethical approval

The Animal Ethics Committee of the College of Veterinary Medicine, Michael Okpara University of Agriculture Umudike approved to this experiment.

2.2 Location of the study

The study was carried out at the Teaching and Research Farm of Michael Okpara University of Agriculture Umudike, Umuahia, Abia State, located within the South East agro-ecological zone of Nigeria.

2.3 Study layout

The research was divided into two studies. Study 1 involved the production of activated charcoal from a blend of locally available agricultural residues such as pig dung, palm kernel shell, and palm fruit fiber and the evaluation of the physicochemical properties. Study 2 involved the evaluation of the effect of dietary inclusion of activated charcoal on growth, carcass yield, hematology, and serum biochemical indices of broiler chickens.

2.4 Collection, drying, and blending of agricultural residues

The palm kernel shell and palm fruit fiber were collected from a palm oil mill while freshly voided pig dung was collected from pig farms using a plastic container. The materials were carefully collected to avoid contamination with sand or other objects. Each material was sun-dried to constant weight and crushed manually using a wooden pestle and mortar. The materials were then blended together at a ratio of 4:3:3 weight for pig dung, palm kernel shell, and palm fruit fiber, respectively, and used to produce the activated charcoal.

2.5 Study 1 (production of activated charcoal)

The physical method of activated charcoal production described by Gunamartha and Widana [30] was employed in the present study. The blended biomass materials were weighed using HN 289 digital scale (Omron Co., Ltd., Japan) and transferred to a clay pot of about 30 liters for carbonization. In addition to contributing to the carbon yield, palm kernel shell and palm fruit fiber also served as combustion accelerants enhancing the pyrolysis of pig dung [31]. The pot containing the precursors was sealed by covering with a metallic lid that had a small vent which limited the entry of oxygen into the mixture. The pot was placed on open fire for a combustion period of 5 hours at which no more smoke was produced from the vent. At this point, water was introduced quickly to stop the carbonization of the biomass and achieve activation. Thereafter, the pot was tightly closed and allowed to cool. The charcoal product was then harvested, rinsed with cold water to remove ash and other debris, dried, and weighed. The dried activated charcoal was transferred to a wooden mortar and ground with pestle into fine powder and stored in an air tight polythene container for

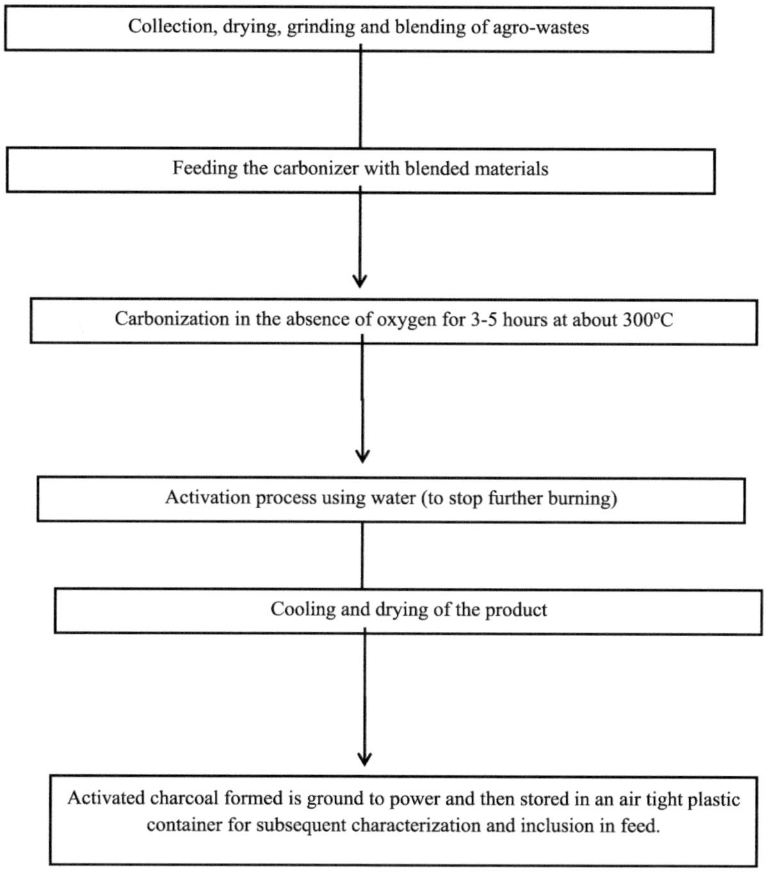

Figure 2.
Flow chat for producing activated charcoal.

characterization and subsequent supplementation in broiler feeds produced according NRC [32] recommendations (**Figure 2**).

2.6 Physicochemical characterization of the activated charcoal

The physical properties determined were bulk density, water holding capacity, specific gravity, moisture content, pH and oil adsorption capacity while the chemical properties were carbon, and mineral contents.

2.6.1 Determination of physical characteristics of activated charcoal

Activated charcoal yield was determined as the ratio of the weight of dried activated charcoal to the weight of precursor carbonized and values were expressed in percentage. The bulk density, water holding capacity, and specific gravity of the activated charcoal were determined according to the procedure described by Makinde and Sonaiya [33] and modified by Omede [34]. The moisture content of the activated charcoal was determined using oven dry method as described by the American Society for Testing and Materials [35] and the percentage moisture content was calculated as recommended by AOAC [36]. The pH of the activated charcoal was determined with the aid of a pH meter (HANNA Combo PH Meter, Model: HI 98129, USA) while oil adsorption capacity was analyzed according to ASTM F 726–99 [37]. The test was performed at 23 ± 4°C with the oil absorbency measured three times and an average value taken according to [38].

2.6.2 Determination of carbon and mineral contents

The concentration of macro minerals namely nitrogen (N_2), calcium (Ca), magnesium (Mg), potassium (K), phosphorus (P), sodium (Na), and micro minerals namely manganese (Mn), zinc (Zn), copper (Cu), and iron (Fe) in the activated charcoal were measured using the Atomic Absorption Spectrophotometer method (Bulk Scientific, 205, India). The procedure was based on the principle that metallic elements in a ground form absorb light of the same wavelength which they emit when excited, with the amount of radiation absorbed being directly proportional to the concentration of the element present.

3. Results

The physical properties of the produced activated charcoal are presented in **Table 1** while the chemical properties are presented in **Table 2**.

The table shows that 7.1 kg activated charcoal was obtained from a total of 10 kg precursor material representing 71.0% activated charcoal yield. Other physical properties were moisture content (5.3%), pH (7.67), bulk density (0.72), water-holding capacity (77.46%), specific gravity (0.73), oil adsorption capacity (118.47%), and surface area (587 cm^2/g).

The carbon content and mineral composition of the activated charcoal produced in this study were as shown in **Table 2**.

Study 2: Evaluation of the effect of dietary inclusion of activated charcoal on growth, carcass yield, hematology, and serum biochemical indices of broiler chickens.

Parameter	Value
Activated charcoal yield (%)	71.0
Moisture (%)	5.37
pH	7.67
Bulk density (g/cm^3)	0.72
Water-holding capacity (%)	77.46
Specific gravity	0.73
Oil adsorption capacity (%)	118.47
Surface area (cm^2/g)	587.00

Table 1.
Physical properties of activated charcoal produced from a blend of agricultural waste materials.

Parameter	Value
Carbon content (%)	79.43
Calcium (mg/kg)	6185.11
Phosphorus (mg/kg)	18,603.29
Sodium (mg/kg)	1722.47
Potassium (mg/kg)	10,275.48
Magnesium (mg/kg)	3980.14
Manganese (mg/kg)	721.00
Iron (mg/kg)	996.35
Zinc (mg/kg)	95.47
Copper (mg/kg)	33.69
Arsenic (mg/kg)	13.38
Nitrogen (mg/kg)	3008.04

Table 2.
Chemical properties of activated charcoal produced from a blend of agricultural waste materials.

3.1 Methodology, experimental birds, and design

One hundred and twenty unsexed day old arbor acre strain of broilers were used. On arrival, they were distributed into four treatment groups (T1–T4) of 30 birds each with each group further replicated three times comprising of 10 birds each in completely randomized design. They were maintained ad-libitum on a starter and finisher diet for 3 weeks each, respectively, with T1 (control) fed diet containing no activated charcoal. T2–T4 was fed diet which contained 0.5 g/kg, 1.0 g/kg and 1.5 g/kg of activated charcoal, respectively. Data collected were feed intake, live weight, carcass weight and organ weight. Live weight gain and feed intake were used to calculate the feed conversion ratio (FCR). Blood samples were collected at the end of 1st, 4th and 6th week of the experiment from the wing vein of the birds into EDTA (Ethylenediamine tetra-acetic acid) and plain bottles for hematological and serum biochemical analysis, respectively. The erythrocyte was counted using the hemocytometer method as describe by Schalm et al. [39] while the hemoglobin concentration was determined according to the techniques described by Cole [40]. In determining the packed cell volume (PCV), the Wintrob microheamatocrit tube method was used while other hematological indices were calculated according to the formula reported by Schalm et al. [39]. Serum biochemical tests were carried out using Randox commercial test kit specific for each biochemical parameter in accordance with standard procedures prescribed by the producer Randox Laboratories (UK). The serum parameters analyzed include the following included total serum protein, serum albumin and globulin, urea, serum creatinine concentration, alanine aminotransferase (ALT), aspartate aminotransferase (AST), alkaline phosphatase activity (ALP), total bilirubin content and total serum cholesterol. Weights of carcass and organ were determined by first slaughtering 12 birds (one per replicate) by severing of the neck with a sharp knife and removal of feathers [41]. Both the carcass and organ weights were expressed as percentage of live weight and data collected were subjected to statistical analyses using ANOVA.

Age	Parameters	Experimental groups			
		T1	T2	T3	T4
Day 1	Live w. (g)	37.80 ± 2.14	38.07 ± 2.41	37.73 ± 2.13	37.43 ± 2.46
Wk 1	Live w. (g)	180.00 ± 36.72	197.33 ± 18.77	219.33 ± 16.26	194.33 ± 11.15
	Weight gain	120.57 ± 1.30	118.63 ± 1.76	122.27 ± 3.25	199.97 ± 5.52
	Feed intake	30.33 ± 2.37^{b}	28.82 ± 2.74^{a}	29.19 ± 1.24^{ab}	28.33 ± 1.72^{a}
	FCR	1.74. ±0.01	1.68 ± 0.07	1.67 ± 0.03	1.65 ± 0.05
Wk 4	Live w (g)	933.00 ± 115.88	1059.67 ± 148.29	1070.67 ± 102.81	1053.33 ± 153.81
	Weight gain	817.37 ± 20.96^{a}	886.26 ± 18.63^{b}	924.87 ± 23.10^{c}	843.27 ± 3.27^{a}
	Feed intake	95.74 ± 8.12^{c}	92.70 ± 7.97^{bc}	89.28 ± 10.34^{ab}	85.93 ± 9.58^{a}
	FCR	1.83 ± 0.05^{d}	1.61 ± 0.03^{c}	1.47 ± 0.03^{a}	1.53 ± 0.02^{b}
Wk 6	Live w. (g)	1947.67 ± 55.08^{a}	2027.67 ± 26.41^{ab}	2114.33 ± 80.21^{b}	1988.33 ± 17.04^{a}
	Weight gain	959.83 ± 54.62	976.67 ± 17.89	1000.33 ± 44.58	964.37 ± 48.34
	Feed intake	156.49 ± 8.00^{b}	150.92 ± 6.72^{a}	150.96 ± 5.67^{a}	153.05 ± 8.00^{ab}
	FCR	2.07 ± 0.07	1.98 ± 0.07	1.94 ± 0.08	2.02 ± 0.11

Table 3.
Growth parameters of broiler chickens fed varying dietary levels activated charcoal.

Tables 3 present the growth performance of the experimental birds while **Table 4** shows the relative organ weights and intestinal parameters of the broiler chicken at 1, 4, and 6 weeks of age.

Results are presented as mean ± standard deviation (n = 3). The results with different letter superscripts are significantly different ($P < 0.05$) from any paired mean across the row.

Results are presented as mean ± standard deviation (n = 3). The results with different letter superscripts are significantly different ($P < 0.05$) from any paired mean across the row.

Tables 5–7 present the hematological indices of the experimental birds at 1, 4, and 6 weeks of age, respectively.

Results in **Table 5–7** are presented as mean ± standard deviation (n = 3). The results with different letter superscripts are significantly different ($P < 0.05$) from any paired mean across the row.

The serum biochemical indices of the experimental birds at week one, four and six are presented in **Tables 8–10**, respectively.

Results of **Tables 8–10** are presented as mean ± standard deviation (n = 30). The results with different letter superscripts are significantly different ($P < 0.05$) from any paired mean across the row.

4. Discussion

4.1 Study 1

The observed activated charcoal yield of 71.0% is less than the 74.19% yield from palm kernel shell alone reported by Kong et al. [42]. The inclusion of pig dung and

Age	Parameters	Experimental groups			
		T1	T2	T3	T4
Wk 1	Live weight (g)	180.00 ± 36.72	197.33 ± 18.77	219.33 ± 16.26	194.33 ± 11.15
	Gizzard+C (%)	7.30 ± 0.60	6.69 ± 0.62	6.82 ± 0.33	7.14 ± 0.41
	Gizzard (%)	4.85 ± 0.22^{b}	4.09 ± 0.19^{a}	4.41 ± 0.13^{a}	4.40 ± 0.27^{a}
	Proventri (%)	1.10 ± 0.18	0.99 ± 025	0.96 ± 0.03	1.08 ± 0.07
	Liver (%)	3.53 ± 0.17	3.37 ± 0.46	3.42 ± 0.11	3.87 ± 0.42
	Heart (%)	0.72 ± 0.10^{a}	0.88 ± 0.12^{ab}	0.91 ± 0.08^{b}	0.84 ± 0.06^{ab}
	Intest. L. (cm)	80.37 ± 3.89^{a}	112.33 ± 3.06^{b}	114.33 ± 9.29^{b}	124.00 ± 3.00^{b}
	Duodenal pH	5.16 ± 0.06	5.30 ± 0.02	5.20 ± 0.17	5.29 ± 0.02
	Jejunum pH	5.16 ± 0.14	5.29 ± 0.01	5.23 ± 0.12	5.29 ± 0.04
	Ileal pH	5.27 ± 0.02	5.30 ± 0.03	5.29 ± 0.03	5.26 ± 0.02
	Rectal pH	5.31 ± 0.02	5.30 ± 0.03	5.30 ± 0.02	5.28 ± 0.04
Wk 4	Live weight (g)	933.00 ± 155.88	1059.67 ± 148.29	1070.67 ± 102.81	1053.33 ± 153.81
	Gizzard+C (%)	3.53 ± 0.22^{a}	3.59 ± 0.27^{a}	4.28 ± 0.15^{b}	4.08 ± 0.14^{b}
	Gizzard (%)	2.47 ± 0.23^{ab}	2.30 ± 0.23^{a}	2.84 ± 0.04^{b}	2.68 ± 0.19^{ab}
	Proventri (%)	0.49 ± 0.06	0.44 ± 0.03	0.53 ± 0.12	0.55 ± 0.01
	Liver (%)	2.52 ± 0.18	2.67 ± 0.33	2.91 ± 0.15	2.84 ± 0.36
	Heart (%)	0.49 ± 0.03	0.53 ± 0.06	0.53 ± 0.10	0.50 ± 0.05
	Intest. L. (cm)	148.00 ± 19.08^{a}	165.00 ± 6.56^{ab}	185.00 ± 23.26^{b}	176.00 ± 13.08^{ab}
	Duodenal pH	6.02 ± 0.82	5.90 ± 0.36	5.93 ± 0.38	6.10 ± 0.26
	Jejunum pH	6.73 ± 0.25^{b}	6.43 ± 0.04^{ab}	6.17 ± 0.15^{ab}	5.29 ± 0.044^{a}
	Ileal pH	7.50 ± 0.50	8.00 ± 1.00	7.17 ± 0.15	7.83 ± 1.36
	Rectal pH	8.10 ± 1.39	8.53 ± 0.90	6.60 ± 0.53	8.50 ± 1.47
Wk 6	Live weight (g)	1947.67 ± 55.08^{a}	2027.67 ± 26.41^{ab}	2114.33 ± 80.21^{b}	1988.33 ± 17.04^{a}
	Carcass. w (g)	1359.64 ± 22.67^{a}	1473.67 ± 34.39^{ab}	1573.33 ± 104.64^{c}	1432.33 ± 22.68^{ab}
	Dressing %	69.81 ± 0.96^{a}	72.67 ± 0.90^{b}	74.38 ± 2.33^{b}	72.03 ± 0.76^{ab}
	Gizzard+C (%)	2.25 ± 0.12	2.30 ± 0.17	2.38 ± 0.11	2.20 ± 0.10
	Gizzard (%)	1.24 ± 0.09^{a}	1.56 ± 0.17^{b}	1.39 ± 0.15^{ab}	1.30 ± 0.06^{a}
	Proventri (%)	0.36 ± 0.02^{a}	0.41 ± 0.02^{b}	0.36 ± 0.01^{a}	0.37 ± 0.02^{a}
	Liver (%)	1.54 ± 0.06	1.67 ± 0.10	1.62 ± 0.11	1.68 ± 0.15
	Heart (%)	0.35 ± 0.02	0.38 ± 0.04	0.39 ± 0.02	0.35 ± 0.02
	Intest. L. (cm)	166.00 ± 17.06^{a}	188.67 ± 9.07^{ab}	207.67 ± 13.80^{b}	187.33 ± 6.51^{ab}
	Duodenal pH	5.93 ± 0.47	6.03 ± 0.64	5.93 ± 0.31	5.87 ± 0.25
	Jejunum pH	6.83 ± 0.12^{c}	6.17 ± 0.21^{b}	6.17 ± 035^{b}	5.97 ± 0.25^{ab}
	Ileal pH	7.60 ± 0.50	7.83 ± 0.55	7.93 ± 0.67	8.20 ± 0.30

Table 4.
Relative organ weight and intestinal parameters of broiler chickens fed varying dietary levels of activated charcoal.

Parameters	T1	T2	T3	T 4
RBC (× 10^6/mm^3)	3.19 ± 0.06^{a}	3.67 ± 0.11^{b}	3.43 ± 0.32^{a,b}	3.61 ± 0.17^{b}
PVC (%)	27.33 ± 0.58^{a}	31.33 ± 1.53^{b}	29.33 ± 2.08^{a,b}	31.67 ± 1.53^{b}
HbC (g/dl)	9.63 ± 0.15^{a}	10.80 ± 0.40^{b}	10.07 ± 0.83^{a,b}	10.87 ± 0.50^{b}
WBC (× 10^3/mm^3)	36.50 ± 0.56^{a}	42.57 ± 3.19^{b}	40.83 ± 4.93^{a,b}	42.97 ± 1.44^{b}
Platelet (× 10^3/mm^3)	158.00 ± 5.29	161.00 ± 11.27	150.67 ± 8.02	160.67 ± 2.52
MCV(fL)	92.96 ± 7.70	90.24 ± 5.99	92.32 ± 5.91	93.07 ± 3.41
MCH (pg/cell)	30.24 ± 0.54	29.37 ± 0.47	29.42 ± 1.29	30.13 ± 0.04
MCHC (g/L)	35.25 ± 0.28	34.48 ± 0.48	33.95 ± 1.13	34.71 ± 1.04
Neutrophil (%)	30.33 ± 1.53	27.67 ± 2.31	29.00 ± 2.65	28.33 ± 0.58
Lymphocytes (%)	63.00 ± 2.00	64.67 ± 1.16	63.67 ± 1.16	65.00 ± 1.73
Monocytes (%)	4.67 ± 0.58	5.67 ± 0.58	5.33 ± 1.53	4.67 ± 1.16
Eosinophils (%)	2.00 ± 0.00	2.00 ± 1.00	2.00 ± 0.00	1.33 ± 0.58
Basophils (%)	0.00 ± 0.00	0.00 ± 0.00	0.00 ± 0.00	0.00 ± 0.00

Table 5.
Hematological indices at 1 week of age for broiler chicks fed varying dietary levels of activated charcoal.

Parameters	T1	T 2	T3	T4
RBC (× 10^6/mm^3)	4.12 ± 0.30^{b}	3.48 ± 0.25^{a}	3.72 ± 0.33^{a,b}	3.37 ± 0.20^{a}
PVC (%)	35.67 ± 2.08^{b}	29.67 ± 2.52^{a}	32.00 ± 2.65^{a,b}	30.67 ± 2.31^{a}
HbC (g/dl)	15.03 ± 0.67^{b}	12.87 ± 0.83^{a}	13.40 ± 1.40^{a,b}	13.17 ± 0.38^{a}
WBC (× 10^3/mm^3)	39.68 ± 2.18	37.43 ± 4.66	39.23 ± 1.36	44.25 ± 5.16
Platelet (× 10^3/mm^3)	270.00 ± 54.53	302.67 ± 10.60	264.33 ± 53.31	295.00 ± 6.00
MCV(fL)	86.80 ± 5.19	85.12 ± 1.81	86.06 ± 0.91	86.74 ± 2.29
MCH (pg/cell)	36.59 ± 1.86	37.18 ± 5.15	35.99 ± 0.58	37.32 ± 1.78
MCHC (g/L)	42.17 ± 0.70	43.72 ± 6.43	41.83 ± 0.90	43.08 ± 3.10
Neutrophil (%)	61.67 ± 1.53	58.33 ± 1.53	61.67 ± 2.08	58.00 ± 3.61
Lymphocytes (%)	30.00 ± 2.00^{a}	36.00 ± 1.73^{b}	29.67 ± 2.52^{a}	33.67 ± 2.52^{a,b}
Monocytes (%)	5.00 ± 1.00^{a,b}	4.33 ± 0.58^{a}	5.67 ± 0.58^{b}	5.00 ± 0.00^{a,b}
Eosinophils (%)	3.33 ± 0.58^{b}	1.33 ± 0.58^{a}	3.00 ± 0.00^{b}	3.33 ± 1.16^{b}
Basophils (%)	0.00 ± 0.00	0.00 ± 0.00	0.00 ± 0.00	0.00 ± 0.00

Table 6.
Hematological indices at 4 weeks of age for broiler chickens fed varying dietary levels of activated charcoal.

palm fruit fiber in the present study may have been responsible for the lower activated charcoal yield obtained. Pig dung and palm fruit fiber are lighter than palm kernel shell and may have lesser carbon contents. The moisture content of the activated charcoal obtained from this study (5.37%) was higher than the 3.43% and 3.50% reported by Okoroigwe et al. [11] and Lima and Marshal [4], respectively, for AC

Parameters	T1	T2	T3	T 4
RBC (× 10^6/mm^3)	4.01 ± 0.17^b	3.79 ± 0.14^{a,b}	3.86 ± 0.04^{a,b}	3.68 ± 0.07^a
PVC (%)	38.00 ± 1.00^c	36.00 ± 1.00^b	36.33 ± 0.58^b	34.33 ± 0.58^a
HbC (g/dl)	14.57 ± 0.06^c	14.03 ± 0.06^b	14.00 ± 0.20^b	13.27 ± 0.25^a
WBC (× 10^3/mm^3)	39.77 ± 1.46	38.50 ± 1.68	38.70 ± 2.55	37.57 ± 1.10
Platelet (× 10^3/mm^3)	254.33 ± 39.72^a	303.00 ± 6.56^b	281.00 ± 11.14^{a,b}	265.67 ± 5.51^{a,b}
MCV(fL)	95.39 ± 2.50	95.02 ± 1.46	94.21 ± 0.68	93.39 ± 1.16
MCH (pg/cell)	36.33 ± 1.43	37.06 ± 1.31	36.30 ± 0.15	36.10 ± 1.37
MCHC (g/L)	3.83 ± 0.09	3.90 ± 0.10	3.85 ± 0.03	3.86 ± 0.14
Neutrophil (%)	53.33 ± 4.93	53.67 ± 3.22	54.67 ± 1.53	55.33 ± 2.08
Lymphocytes (%)	39.33 ± 4.73	39.33 ± 4.04	37.67 ± 2.08	37.33 ± 2.31
Monocytes (%)	4.67 ± 0.58	4.00 ± 0.00	4.67 ± 0.58	4.00 ± 0.00
Eosinophils (%)	2.67 ± 0.58	3.00 ± 1.00	3.00 ± 0.00	3.33 ± 0.58
Basophils (%)	0.00 ± 0.00	0.00 ± 0.00	0.00 ± 0.00	0.00 ± 0.00

Table 7.
Hematological indices at 6 weeks of age for broiler chickens fed varying dietary levels of activated charcoal.

Parameters	T 1	T2	T3	T 4
Total protein (g/dl)	2.45 ± 0.15^a	3.03 ± 0.34^c	2.56 ± 0.22^{ab}	2.97 ± 0.13^{bc}
Albumin (g/dl)	1.36 ± 0.04^a	1.52 ± 0.05^b	1.28 ± 0.02^a	1.55 ± 0.14^b
Globulin (g/dl)	1.09 ± 0.15	1.44 ± 0.27	1.28 ± 0.20	1.42 ± 0.17
AST (m/l)	36.00 ± 6.00	40.00 ± 2.65	39.33 ± 5.13	33.33 ± 2.89
ALT (m/l)	23.33 ± 2.31^{ab}	27.33 ± 2.52^b	23.67 ± 2.31^{ab}	21.67 ± 2.52^a
ALP (m/l)	88.67 ± 5.51^a	90.00 ± 2.00^{ab}	97.67 ± 3.79^{bc}	99.00 ± 5.29^c
Bilirubin (mg/l)	0.45 ± 0.05	0.49 ± 0.05	0.43 ± 0.02	0.48 ± 0.07
Cholesterol (mg/dl)	83.03 ± 2.31^a	105.11 ± 1.12^b	90.42 ± 2.66^a	85.28 ± 6.67^a
Urea (mg/dl)	9.37 ± 0.66	9.93 ± 2.33	9.44 ± 0.73	9.03 ± 0.42
Creatinine (mg/dl)	0.88 ± 0.07	0.80 ± 0.08	0.87 ± 0.06	0.79 ± 0.05

Table 8.
Serum biochemical indices at 1 week of age for broiler chicks fed varying dietary levels of activated charcoal.

derived from palm fruit fiber, and poultry litter, respectively. It was also higher than that of Nwankwo [43] who recorded 3.50% as moisture content of AC from cow bone sourced from abattoirs. Kong et al. [42] investigated the moisture content of activated charcoal derived from PKS and reported a lower moisture content of 1.47%. The lower moisture contents of the AC reported by these researchers could be attributed to higher carbonization and activation temperatures which were over 600°C in each case as against the 300°C [44] employed in the present study. The AC exhibited a slightly alkaline pH value of 7.67 which was within the preferred range of 6.0–10.0 as reported for most agricultural residue-derived AC [45]. It has also been noted that acidic and slightly alkaline activated charcoals exhibited greater adsorption capacity and are

Parameters	T1	T2	T3	T 4
Total protein (g/dl)	3.27 ± 0.24	3.26 ± 0.21	3.18 ± 0.28	3.14 ± 0.13
Albumin (g/dl)	1.90 ± 0.06	2.09 ± 0.17	2.02 ± 0.35	1.90 ± 0.06
Globulin (g/dl)	1.37 ± 0.27	1.17 ± 0.06	1.16 ± 0.10	1.23 ± 0.08
AST (m/l)	36.33 ± 3.21	38.67 ± 5.13	38.67 ± 2.08	35.33 ± 5.03
ALT (m/l)	30.67 ± 1.15	28.00 ± 3.00	25.33 ± 2.52	26.00 ± 6.25
ALP (m/l)	91.07 ± 5.00	91.60 ± 4.44	92.80 ± 3.22	93.13 ± 3.07
Bilirubin (mg/l)	0.53 ± 0.06	0.52 ± 0.09	0.57 ± 0.04	0.46 ± 0.15
Cholesterol (mg/dl)	103.37 ± 3.84[c]	96.27 ± 0.71[b,c]	92.03 ± 2.37[a,b]	84.43 ± 9.04[a]
Urea (mg/dl)	12.20 ± 1.87	14.80 ± 0.53	14.60 ± 2.23	13.33 ± 1.59
Creatinine (mg/dl)	0.79 ± 0.08	0.79 ± 0.09	0.87 ± 0.15	0.87 ± 0.17

Table 9.
Serum biochemical indices at 4 weeks of age for broiler chicks fed varying dietary levels of activated charcoal.

Parameters	T1	T2	T3	T4
Total protein (g/dl)	3.46 ± 0.03[c]	3.21 ± 0.03[b]	3.12 ± 0.08[a,b]	3.06 ± 0.07[a]
Albumin (g/dl)	2.18 ± 0.08[c]	1.92 ± 0.04[b]	1.80 ± 0.01[a]	1.80 ± 0.04[a]
Globulin (g/dl)	1.28 ± 0.10	1.29 ± 0.03	1.32 ± 0.08	1.26 ± 0.03
AST (m/l)	40.33 ± 0.58[b]	36.67 ± 2.89[a,b]	34.33 ± 2.08[a]	40.33 ± 1.53[b]
ALT (m/l)	29.67 ± 1.53	27.33 ± 2.52	27.00 ± 2.65	28.67 ± 1.15
ALP (m/l)	85.67 ± 4.04	89.00 ± 3.61	83.33 ± 3.06	84.67 ± 2.52
Bilirubin (mg/l)	0.56 ± 0.04	0.60 ± 0.03	0.58 ± 0.02	0.57 ± 0.03
Cholesterol (mg/dl)	97.17 ± 0.95[c]	90.87 ± 1.32[b]	86.33 ± 5.05[a,b]	81.50 ± 0.92[a]
Urea (mg/dl)	9.50 ± 0.28	9.57 ± 0.99	9.17 ± 0.13	9.07 ± 0.21
Creatinine (mg/dl)	0.84 ± 0.02	0.81 ± 0.04	0.83 ± 0.02	0.86 ± 0.06

Table 10.
Serum biochemical indices at 6 weeks of age for broiler chicks fed varying dietary levels of activated charcoal.

more effective adsorbents when compared to those with very high pH values [46]. The pH result obtain from this study was however higher than the 6.1, 6.64, and 6.60 reported by Okoroigwe et al. [11], Evbuoman et al. [47] and Nwankwo [43], respectively, using agricultural residues as precursors for pyrolysis. The observed bulk density of 0.72 g/cm^3 was higher than 0.49 g/cm^3 reported by Evbuoman et al., [47] for PKS derived-activated charcoal. The value of the bulk density is however within the preferred range of 0.06–1.03 g/cm^3 as recommended by Bryne and Nagle [48] for activated charcoal with high adsorption capacity and micro-porosity. The water-holding capacity of 77.46% reported in this experiment was higher than the value of 47.4% obtained by Kong et al. [42] for PKS derived activated charcoal. Mollinedo et al. [49] demonstrated the use of AC to improve the water-holding capacity of different soil samples and discovered that treatment of soil increased water retention capacity by 25% when compared with untreated control.

The AC produced in this study is suitable for increasing the water retention capacity of soil considering its high water-holding capacity. Enhanced soil water retention will improve plant nutrient availability and uptake; thereby improving crop yield [50, 51]. It will have additional fertilizer value because of its high concentration of important plant macro nutrients such as potassium and phosphorus [52]. The use of rice husk-activated charcoal to fertilize rice fields had been a common practice in Asian countries [52]. The urban encroachments on poultry facilities have resulted in increased complaints from local residents [53, 54], due to bad odor and nuisance flies. In addition, farmers incure huge economic losses associated with poor litter in poultry farms resulting from foot and leg problems, respiratory diseases, poor weight gain and inferior feed conversion [55]. The high water holding capacity of AC could be beneficial in minimizing problems associated with wet litter in livestock and poultry farms. Sashikala et al. [56] compared the odor abatement of poultry litter using three odor control products (activated charcoal, silica gel, and zeolite) under controlled environmental conditions and reported that activated charcoal and silica gel exhibited prominent adsorption or reduction in litter volatiles. Specific gravity (SG) otherwise called relative density is the ratio of the density of substances to the density of water [34]. This physical parameter plays a vital role in the transit of digesta through the gastrointestinal (GIT) tract of animals [57]. The value of the specific gravity obtained in this research 0.730 was lower than the 1.61 reported by Evbuoman et al. [47] but higher than the 0.64 reported by Okoroigwe et al. [11] for bamboo and palm kernel shell-derived activated charcoal, respectively. It should be recalled that particles with specific gravity of less than 1.20 were more likely to float in the gastrointestinal tract of animals thereby increasing their retention time while those greater than 1.50 sink leading to a reduced retention time [57, 58]. The specific gravity recorded in this study was far higher than the range of 0.33–0.46 reported by Omede [34] for conventional feed ingredients produced in Nigeria and hence may enhance the specific gravity of feeds when supplemented in rations.

The oil adsorption capacity (OAC) and surface area (SA) obtained in this experiment were 118.47% and 587cm^2/g, respectively. The value for surface area is higher than the range of 248–253 cm^2/g reported by Lima and Marshal [4] for AC derived from poultry litter material. The high surface area coupled with the slightly alkaline pH of 7.67 could be responsible for the high oil adsorption capacity of 118.42% observed. It has been reported that low pH values and high surface area tend to increase the oil adsorption capacity of ACs [46]. With these outstanding properties, the activated charcoal derived from this study could be beneficial for gastrointestinal de-contamination when used as feed additive [55, 59–61]. It has also been reported that low-cost materials such as palm kernel shell, palm fruit fiber, and animal wastes are good precursors for producing AC for use as adsorbents because of well-developed pore structure and high surface area responsible for extensive adsorption capacity [62, 63]. Therefore, the AC produced in this study could be suitable for use in water remediation in cases of oil spillage in oil producing communities [60]. Activated charcoal produced from readily available and renewable agricultural residues would be less expensive and serve as replacement for other more costly adsorbents imported for this and similar purposes, thereby transforming waste into wealth [5, 64]. Furthermore, natural water sources available to most communities in developing countries like Nigeria are rivers, and natural ponds mostly contaminated with heavy metals and effluents discharged from industries [61]. Studies by [65, 66] showed that such heavy metal contaminated water used in animal feeding have negative effects on performance. Activated charcoal such as produced in the present study could be suitable for purifying contaminated water for farm and domestic use by adsorption of metallic ions and bacterial toxins [59].

The value of the carbon content was 79.43% which is higher than the 65.4% reported by [42]. It is also higher than that of wood-derived activated charcoal (AC) (71.40%) and coconut shell-derived AC (60.07%) as reported by Widowati and Asnah [67] but lower than the 85.0, and 88.4% reported by Hidayu and Musa [68], and Okoroigwe et al. [11], respectively, using palm kernel shell and oil palm fiber as precursor materials. Lima and Marshal [4] pyrolysed poultry litter and recorded a carbon content of 29% which was far below the carbon content obtained in this study. The value obtained in the present experiment was however within the preferred range of 62.20–92.40% recommended by Domingues et al. [69] for activated charcoal with high degree of micro-porosity and adsorption capacity. More so, the carbon content value obtained in this study can be adjudged to be high when compared to the International Biochar Initiative (IBI) standard which requires 10% minimum organic carbon in activated charcoal [70]. The European Biochar Foundation also recommended that for any residue left after pyrolysis to qualify as activated charcoal, the carbon content should not be less than 10% [71]. Several studies have shown that the most important factors that affect carbon yield and carbon content of AC are density and nature of the carbonized material or precursor [72]. This could be the reason why different agricultural residues exhibit different physicochemical characteristics even with the same method of treatment or activation. Martinez et al. [73] observed that the texture, carbon yield and carbon content as well as development of pores of AC were strongly affected by the physical and chemical characteristics of the starting material or precursor.

The concentration of minerals evaluated in the present study were calcium (6185.11 mg/kg), phosphorus (18,603.29 mg/kg), sodium (1722.47 mg/kg), potassium (10,275.48 mg/kg), magnesium (3980.14 mg/kg), manganese (721.00 mg/kg), iron (996.35 mg/kg), zinc (95.47 mg/kg), copper (33.69 mg/kg), arsenic (13.38 mg/kg), and nitrogen (3008.04 mg/kg). These mineral concentrations were much higher than the values reported by Okoroigwe et al. [11] and that of Gunamartha and Widana [30] for PKS and cow dung-derived activated charcoals, respectively. These variations could be attributed to the nature of the starting material (precursor) which influences the mineral composition and concentration of the resulting activated charcoal [74–76]. More so, the properties of the AC and its elemental composition can be influenced by the method of activation, duration of activation, and carbonization temperature [77, 78]. The high concentration of potassium in the activated charcoal produced in this experiment could be attributed to the inclusion of palm fruit fiber as one of the precursors. Activated charcoal rich in potassium could serve as fertilizers to enrich soils for enhanced crop yield [79]. The heavy metals, for example, arsenic and the micro mineral (zinc) were within the allowable threshold for these elements in activated charcoals namely lead <150 mg/kg, copper <30 mg/kg, zinc <400 mg/kg, and arsenic <30 mg/kg [70].

4.2 Study 2

The feed intakes of broilers as seen in **Table 3** were higher in the control when compared with the supplemented groups except in week 5 and in week 3 where that of the control did not vary significantly with group 2 probably because of its lowest inclusion rate. This trend of reduction in feed intake in the supplemented groups (T2–T4) was in agreement with the report of Kutlu et al. [80] who observed that activated charcoal reduced feed intake which was attributed to higher bulk density of activated charcoal supplemented feeds [81]. More so, the blackening of feeds by charcoal might cause reduction in palatability [81, 82] which could be responsible for the significant reduction in feed intakes for the supplemented broiler groups. The implication is that feedstuffs with high bulk density exhibit high water-holding capacity and absorb excess water in the gastrointestinal tract (GIT) capable of triggering satiety resulting to low feed intakes [34]. Satiety signal such as cholecystokinin (CCK) provides information about feed intake to the brain which thereby suppress appetite [83–86]. As for the control group, there was a decreased expression of the satiety receptor (CCKR) that was responsible for the increased feed intakes [87]. These reductions in feed intake in the supplemented groups is expected in view of the fact that activated charcoal is a prebiotic which improve the nutrient status of animal by enabling more and efficient use of the nutrient present in the diet and not by stimulating appetite [88]. The non-significant variation in feed intake between broilers in group 1 and group 2 at 3 weeks of age may be attributed to the lowest inclusion of activated charcoal (0.5 kg/100 kg of feed) in diet of broilers in group T2 which may not have adversely affected feed intake. More so, the blackening of the feeds by charcoal at this lowest inclusion was also not too noticeable.

According to the results in **Table 3**, the live weight of broilers in the supplemented groups (T2–T4) was significantly higher than the control group (T1) at 6 weeks of age with the exception of group 4 that did not vary significantly with group 1. This non-significant variation between the live weights of broilers in groups T1 and T4 can be attributed to the highest inclusion level of activated charcoal in the diet of broilers in group 4 which was 1.5 kg/100 kg of feed which maximally reduced feed intake. The increment in the live weights and weight gains in group T2 and T3 as shown in **Table 3** and the FCR that favored the supplemented groups were in agreement with the findings of Dim et al. [26] who reported that the final body weight, average daily weights and FCR favored birds placed on diet supplemented with activated charcoal than the control after 56 days trial period. The results were also in conformity with the report of Jiya et al. [24] whose results showed improved performance on inclusion of activated charcoal in broiler diets. The effect on live weights and weight gains were significantly better in group 3 with 10% charcoal inclusion than in group T2 and T4. These results were exactly similar with the findings of Durunna et al. [25] who recommended 1.0 kg/100 kg of the feed as the best inclusion level for broilers as against 0.6 kg/100 kg of feed by Dim et al. [26]. The live weights of broilers in group T4 were comparable to those in group T1 probably due to the high bulk density and the blackening of the broiler diets of G4 at 1.5 kg/100 kg inclusion that resulted to the lower feed intakes [80–82]. According to the results in **Table 3**, the feed conversion ratio (FCR) were better in group T3 than other groups especially at week 4 where it showed significant difference. The improvement in the feed conversion efficiency in the supplemented groups especially in group T3 could be attributed to the ability of the birds fed AC to maximally utilize the vitamin-mineral premix especially iron and B-complex vitamins in the diet probably due to the binding of AC with toxins and anti-nutritional factors in the gut [26].

More so, the higher intestinal length of the supplemented groups as seen in **Table 4** could be responsible for their high performance due to increased area available for adsorption of nutrients coupled with the significant reductions in the pH of their jejunum at 4 and 6 weeks of age. The weight of the gizzard with content and gizzard were significantly higher in group T3 and T2, respectively, at 4th and 6th week in each case which resulted to their increased dressing percentages. The weight of the proventriculus and heart were significantly higher in groups T2 and T3, respectively, which together with gizzard are the major organs determining performance of birds and the economics of production [89]. Weight of internal organs expressed as percentage of live weight were significantly higher with respect to gizzard and gizzard with content in groups T3 and T4 at 4 weeks of age than group 2 while the gizzard of group T2 and T3 were significantly higher than the control at 6 weeks of age as shown in **Table 4** as against the gizzard of the control group that was relatively higher at first week of age. More so, the weight of the heart expressed as percentage of live weight was significantly greater in the supplemented groups than the control group T1. These significant variations in weights are expected as the gut (gizzard) and heart are the major organs determining performance of birds and the economics of production [89]. The significant differences noticed in the weights of the heart in relation to live weight of the supplemented group can be attributed to the lowering effect of AC on serum cholesterol levels which may be responsible for the increased activity and weights of the heart. Activated charcoal interferes with the entero-hepatic circulation of bile acid and cholesterol, thereby lowering serum cholesterol in cases of hypercholesterolemia [90, 91]. In a related development, Shabani et al. [92] and Dim et al. [26] reported that plasma cholesterol levels were reduced in birds whose diets were supplemented with activated charcoal.

According to **Table 4**, there were no significant difference in the liver and proventricular weights relative to live weight between the broilers in the supplemented group and the control except at 6th week where the relative weights of the proventriculus to live weight was significantly higher in T2 than other groups. This was in agreement with the findings of Majewska and Zaborowski [93] that liver weights did not show any significant variations between the groups whose diets were supplemented with AC and control. There were no significant difference in the weights of the gizzard + contents relative to live weight between the groups at first and 6 weeks of age except at 4 weeks of age when groups T3 and T4 were significantly higher than group T1 and group T2. At 4 weeks of age, the relative weight of the gizzard to live weight was significantly higher in group 3 than other groups and at 6 weeks of age by group T3 and group T2. This higher relative weight of the gizzards to live weights in group T3 in the 4th and 6th week could be responsible for its higher performance since weight is an index of growth and performance.

At 1 week of age, hematological values (RBC, PCV, HbC, and WBC) were significantly higher in the supplemented treatment groups (T2 and T4) than the control (T1). These results were in conformity with the report of Dim et al. [26] who reported that AC inclusion in broiler feeds improved significantly the hematological indices such as hemoglobin concentration (Hb) and red blood cells counts (RBC). This trend was not the case at 4 and 6 weeks of age where T1 recorded RBC, PCV, and Hb that were significantly higher than T4. At 1 week of age, the broilers in the supplemented treatment groups (T2–T4) had better hematological picture than the control but with the continuous supplementation of activated charcoal till 6th week, it turned in favor of T1 followed by T3 signifying a negative correlation.

In accordance with Jindal et al. [82] and Evans et al. [81], the inclusion of AC in poultry feeds increased the bulk density and caused blackening of the feeds which caused some degree of unpalatability responsible for low feed intakes and subsequent reduction RBC counts, PCV and Hb especially in T4 with the highest inclusion level of AC as witnessed in the 4th and 6th week of age. The hematological components which consist of PCV, RBC, Hb, MCV, MCH and MCHC all fell within the normal range for broilers as reported by previous researchers with MCV, MCH and MCHC showing no significant differences between the supplemented treatment groups (T2–T4) and the control (T1) as shown in **Tables 1** and **3**. Iyaode et al. [94] reported normal hematological range for broilers as 25.60–32.50% for PCV, 8.93–10.45 g/dl for Hb and 3.53–3.80 x 10^6/μL for RBC count in broilers. Marcos et al. [95] reported broiler hematological references range to be 22–35% for PCV, 2.5–3.5 x 10^6 / μL for RBC, 7–13 g/dl for Hb and 12–30 × 10^3/μL for WBC. Hidayat et al. [96] recorded normal range for Hb in broilers as 6.65–7.4 g/dl while Salam et al. [97] and Sugiharto et al. [98] both reported range of hemoglobin concentration (Hb) in broilers to be between 5.18 and 9.30 g/dl.

The results were in agreement with the report of Dim et al. [26] noted that the hemoglobin concentration (Hb) and the red blood cell (RBC) count were significantly improved, while the cholesterol levels were significantly reduced in broilers whose diets were supplemented with activated charcoal. The authors attributed the ability of the birds fed activated charcoal to maximally utilize the vitamin-mineral premix in the diet especially iron and B-complex vitamins probably due to the binding of activated charcoal with toxins and anti-nutritional factors in the gut of bird. At 4 and 6 weeks of age, there were dose-dependent reductions in the serum cholesterol levels in the supplemented treatment groups (T2–T4) which were significantly lower when compared with T1 as shown in **Tables 8–10**. This was in agreement with the results of previous researchers that confirmed that serum cholesterol levels were reduced in birds whose diets were supplemented with activated charcoal [26, 92].

These reductions in serum cholesterol levels have elucidated the fact that activated charcoal could be useful in the treatment of hypercholesterolemia [90, 91]. This is achieved by its interference on entero-hepatic circulation by binding to cholesterol and cholesterol-containing bile acids in the gut, thus preventing them from being absorbed [99]. When bile acids are excreted, plasma cholesterol is converted to bile acids to normalize bile acids levels which eventually lowers plasma and serum cholesterol levels [90, 91]. It should be recalled that approximately 2/3 of intestinal cholesterol is derived from bile while just about 1/3 comes from diet [91, 99]. Hence the serum cholesterol is determined by the balance of its synthesis, catabolism and intestinal absorption [90]. The most common statin based therapy for hypercholesterolemia acts by inhibiting the HMG-CoA reductase enzyme to reduce cholesterol synthesis while activated charcoals are useful in reducing the intestinal absorption [90, 91]. Therefore, blocking intestinal absorption is a key point in hypercholesterolemia therapy. Activated charcoal used in this study was able to absorb excess of cholesterol in the intestine before it entered the blood circulation as corroborated by the reports of Joseph et al. [99] and Roosdiana et al. [91]. This is in agreement with the findings of Boonanuntansarn [100] that activated charcoal had a significant reduction on the blood cholesterol levels in 4-week old Nile Tilapia.

Ugbogo et al. [101] stated that although cholesterol plays central role in many biochemical processes where it helps to digest fats, strengthen cell membranes and make hormones, it is majorly known for its association with cardiovascular diseases. Hypercholesterolemia is a metabolic disease which is caused by an elevated total cholesterol level in blood circulation. This may result to its build up on arterial walls, hence narrowing the lumen and increasing risk of blood clots, heart attack and stroke often associated with disease like diabetes mellitus, hypertension and some form of thyroid, liver and kidney disease [91, 101]. About 40–70% of the world's population suffers from hypercholesterolemia [102] and it had also been reported in cats and dogs [103]. A long-term treatment of this condition using a synthetic drug known as statin was associated with side effect including joint pains and liver damage [91]. It became imperative to explore alternative medication derived from natural products such as AC to overcome this problem.

At 1 week of age, the ALP, albumin, and total proteins were significantly higher in broiler groups whose diets were supplemented with activated charcoal (T2–T4). This was unlike at 6 weeks of age where the total protein and albumin level were significantly higher in T1 than in T2, T3 and T4 with its AST significantly higher than T3. According to Ugboho et al. [101], total protein and ALP levels present in the blood can be used to evaluate unexplained weight loss and symptoms of liver damage. The ALP, albumin and total protein were significantly higher in the supplemented treatment groups but the trend was not maintained at 6 weeks of age. The significant increase in the albumin and total protein levels in the control (T1) at 6 weeks of age when compared to the supplemented treatment groups (T2–T4) can be due to the metabolic demand from the liver resulting from the high feed intake [94]. Most of the biochemical parameters were within the patterns often found in avian species as reported by Marcos et al. [95]. The total proteins recorded in this study were in the range of 2.5–4.5 g/dl as cited by Thrall [104] while the ALT were in range of 19–50 μ/L as reported by Lumeji [105]. Globulin also was within the normal range for Gallus gallus specie (0.5–1.8 g/dl) as reported by Thrall [104].

5. Conclusion and recommendation

The study showed that the activated charcoal produced using these agricultural residues (pig dung, palm fruit fiber, and PKS) was of high physicochemical properties within the range of most activated charcoals produced for gastrointestinal decontamination, water treatment and environmental remediation. Its inclusion in broiler feed improved performance and carcass yield and could serve as alternative feed additive in view of the ban placed on sub-therapeutic inclusion of antibiotics for growth promotion due to antimicrobial resistance. The hematological and biochemical parameters examined were within the patterns often found in avian species signifying that activated charcoal is non-toxic and safe to be used in oral administration at best inclusion level of 1 kg/100 kg of feed. Therefore, the agricultural waste-derived activated charcoal used in this study is suitable for improvement of hematological parameters in young chicks and in cases of hypercholesterolemia to bind cholesterol and cholesterol-containing bile acids in the gut. It could also serve as a replacement for the synthetic drug used for this condition which is currently very expensive coupled with their long standing side effects which have generated a lot of complaints from patients.

Acknowledgements

The authors wish to express their gratitude to Prof. Charles I. Okoli of the Department of Animal Science and Technology, Federal University of Technology Owerri, Nigeria for his effort in correcting the manuscript and for providing all the technical supports.

Conflict of interest

The authors wish to declare that there is no conflict of interest whatsoever in this chapter contribution.

Author details

Stephen Nnaemeka Okey* and Cosmas Chikezie Ogbu
Department of Veterinary Biochemistry and Animal Production, College of Veterinary Medicine, Michael Okpara University of Agriculture Umudike, Umuahia, Nigeria

*Address all correspondence to: nnamuokey@gmail.com

References

[1] Makkar HPS, Beever P. Optimization of feed use efficiency in ruminant production systems. In: Proceedings of the FAO Symposium; 27 November 2013; Bangkok, Thailand. 2013

[2] Obi FO, Ugwuishiwu BO, Nwakaire JN. Agricultural waste concept, generation, utilization and management. Nigerian Journal of Technology. 2016;**35** (4):957-964

[3] Balat M, Kirtary E, Balat H. Main routes for the thermo-conversion of biomass into fuels and chemical. Part 1: Pyrolysis systems. Energy Conversion and Management. 2009;**50**:3147-3157

[4] Lima IM, Marshal WE. Granular activated carbons from broiler manure: physical, chemical and adsorptive properties. Journal of Bioresource Technology. 2004;**96**:699-709

[5] Ahmedna M, Marshal WE, Rao RM. Production of granular activated carbons from selected agricultural by-products and evaluation of their physicochemical and adsorption properties. Bioresource Technology. 2000;**71**:113-123

[6] Sugumauan P, Susan VP, Ravichandran P, Seshadri S. Production and characterization of activated carbon from banana empty fruit bunch and *Delonix regina* fruit production. Journal of Sustainable Energy and Environment. 2012;**3**:125-132

[7] Agba MM, Ushie EF, Abam I, Agba MS, Okoro J. Developing the biofuel industry for effective rural transportation. European Journal of Scientific Research. 2010;**40**:441-449

[8] Akorode MF, Ibrahim O, Amuda SA, Otuoze AO, Olufeagba BJ. Current status and outlook of renewable energy development in Nigeria. Nigerian Journal of Technology. 2017;**36**(6): 196-212

[9] Mohammed YS, Mustafa MW, Bashir N, Mokhatar AS. Renewable energy resources for distributed power generation in Nigeria: A review of the potential. Renewable and Sustainable Energy Reviews. 2013;**22**: 257-268

[10] Zafar, S. (2018). Palm Kernel shell as biomass resource. Available from: http://www.bioenergyconsult.com/tag/comb ustion. pp. 12-13

[11] Okoroigwe EC, Ofomatah AC, Oparaku NF, Unachukwu GO. Production and evaluation of activated carbon from palm kernel shell for economic and environmental sustainability. International Journal of Physical Science. 2013;**8**(19):1036-1041

[12] Okoli, I.C. (2020). Agricultural residues: Abandoned wealth being recovered by tropical research. Available from: http://researchtropica.com/agric ultural-residues-abandoned-wealth-be ing-recovered-by-tropical-research/ [Accessed June 22, 2020]

[13] Dhyani V, Bhaskar T. A comprehensive review on the pyrolysis of lignocellulosic biomass. Renewable Energy. 2017;**4**:35

[14] AAFCO. Official Publication. USA: Association of American Feed Control Officials; 2012

[15] Addul A, Aberuagba F. Comparative study of the adsorption of phosphate by activated carbon from corncobs, groundnut shell and rice husk. AUJ Journal. 2005;**9**(1):59-63

[16] Schmidt H, Hagemann N, Draper K, Kammann C. The use of biochar in animal feeding. Peer Journal. 2019;7: e7373. DOI: 10.717/peerj.7373

[17] Khan AM, Ansari R. Activated carbon preparation, characterization and applications: A review article. International Journal of Chemical Technology Resource. 2009;**1**(4):859-864

[18] Bacaoui A, Yaacoubi A, Dahbi A, Bennouna C, Phan Tan Luu R, Maldonado-Hodar FJ, et al. Optimization of conditions for the preparation of activated carbons from olive-waste cakes. Carbon. 2001;**39**:425-432

[19] Chyka PA, Seger D, Krenzelok EP, Vale JA. Single-dose activated charcoal. Clinical Toxicology. 2005;**43**(2):61-87

[20] American Academy of Clinical Toxicology, AACT. Position statement and practical guidelines on the use of multi-dose activated charcoal in the treatment of acute poisoning. American Academy of Clinical Toxicology. Journal of Toxicology. Clinical Toxicology. 1999; **37**:731-751

[21] Davis. Atherosclerosis an inflammatory process. Journal of Insurance Medicine. 2005;**37**:72

[22] Majewska T, Pudyszak K, Koziowski K. The effect of charcoal addition to diets for broiler performance and carcass parameters. Veterinary Medicine Zootechnika. 2011;**55**(77):30-32

[23] Bhatti SA, Khan MZ, Hassan ZU, Saleemi MK, Saquib M, Khatoon A, et al. Comparative efficacy of bentolite and activated charcoal in regulating feed transfer of mycotoxins. Journal of the Science of Food and Agriculture. 2018; **98**(3):888-890

[24] Jiya EZ, Anyanwale BA, Ijaiya AT, Ugochukwu A, Tsado D. Effect of activated coconut shell charcoal meal on growth performance and nutrient digestibility of broiler chicken. British Journal of Applied Science and Technology. 2013;**3**(2):268-276

[25] Durunna, C.S., Abatai, U.G and Uchegbu, C. (2018). Performance evaluation of broiler chickens fed varying levels of raw African velvet tamarind, Icheku (*Dialium guineense*) wood charcoal as feed additive. Proceedings of 43rdAnnual Conference of the Nigerian Society for Animal Production, 18–22 March 2018; Owerri, Imo State, Nigeria. pp. 514-516

[26] Dim CE, Akuru EA, Egom MA, Nnajofor NW, Ossai OK, Ukaigwe CG, et al. Effect of dietary inclusion of biochar on growth performance, haematology and serum lipid profile of broiler birds. Agro Science. 2018; **17**(2):8-16

[27] Moyes CD, Schute PM. Principles of Animal Physiology. 2nd ed. New York, NY: Pearson International Edition; 2008

[28] Sufiriyanto NI, Emmy S. Hematological profiles and performance of broiler chickens fed on commercial feed. Journal of Animal Production. 2018;**20**(3):183-190

[29] Bawala, T. O., Akpan, U., Ogunnowo, A.O., Fusae, O. A. and Sogunle, O. M. (2007). The influence of magnesium supplementation on the hematological profile of young West Africa dwarf goats. Proceedings of the 32nd conference of the Nigerian Society for Animal Production, 76-78

[30] Gunamantha IM, Widana GAB. Characterization of the potential of biochar from cow and pig manure for genecology application. Conference Series: Earth Environmental Science. 2018;**131**:12-55

[31] Nwokolo, C. O. and Ogunyemi, S. (2008). Empirical Study on Optimizing Recovery from Oil Palm Waste in Nigeria. In: International Conference on Renewable and Alternative Energy; March 29-April 2, 2009; Owerri, Nigeria. Books of Abstracts. p. 24

[32] NRC. Nutrient Requirement of Poultry. Washington, DC: National Research Council National Academic Press; 1994

[33] Makinde, O. A. and Sonaiya, E. B (2007). Determination of water holding capacity, blood and rumen fluid adsorbance of some fibrous feed stuff. In: A. Giang et al., editors. Sustainability of Livestock Industry in Oil Economy. Proceedings of the 32nd Annual Conference of the Nigeria Society for Animal Production; 28: 84-87

[34] Omede, A. A. (2010). The use of physical characteristics in the quality evaluation of commercial feeds and feedstuffs [MSc thesis]. Federal university of Technology, Owerri, Nigeria

[35] ASTM D 280-33. Standard Test Methods for the Hydroscopic Moisture Content of Activated Carbon. West Conshohocken, PA: ASTM; 2003

[36] AOAC. The Official Methods of Analysis. 13th ed. Washington, DC: Association of Official Analytical Chemists; 1990

[37] ASTM F 726-99. Standard Test Methods for Oil Adsorption Capacity of Activated Carbon. West Conshohocken, PA: ASTM International; 1998

[38] Hussein M, Amer AA, Sawsan II. Oil spill adsorption using carbonized pith bagasse: Trial for practical application. International Journal of Environmental Science and Technology. 2008;**5**:233-242

[39] Schalm OW, Jain NC, Carrol EJ. Veterinary Haematology. 3rd ed. Philadelphia, PA: Lea and Febiger; 1975. pp. 197-199

[40] Coles EH. Avian Clinica Pathology. Philadelphia, PA: W.B. Sanders Company; 1986

[41] Scott ML, Neshein MC, Young RJ. Nutrition of Chicken. 1st ed. Ithaca, NY: Scott and Associates; 1969

[42] Kong SH, Loh SK, Bachmann RJ, Choo YM, Abdu Rahim S. Production and physico-chemical characterization of biochar from palm kernel shell. Food Science and Technology Postgraduate Colloquium, AIP Conference Proceeding. 2013;**1571**:749-752

[43] Nwankwo IH. Production and characterization of activated carbon from animal bone. American Journal of Engineering Research. 2018;7(7):335-341

[44] Prakash Kumar BG, Shivakamy K, Miranda LR, Velan M. Preparation of steam activated carbon from rubber and wood sawdust *(Heveabrailienis)* and its adsorption kinetics. Journal of Hazardous Materials B. 2006;**136**:922-929

[45] Chen X, Jeyaseelam S, Graham N. Physical and chemical properties of activated carbon made from sewage sludge. Journal of Waste Management. 2002;**22**:755-760

[46] Madu PC, Lajide, l. Physicochemical characteristics of activated charcoal derived from melon seed husk. Journal of Chemical and Pharmaceutical Research. 2013;**5**(5):94-98

[47] Evbuomwan BO, Abutu AS, Ezeh CP. The effects of carbonization temperature on some physicochemical properties of bamboo based activated carbon by potassium hydroxide (KOH)

activation. Greener Journal of Physical Science. 2013;**3**(5):187-191

[48] Byrne CE, Nagle DC. Carbonization of wood for advanced materials applications. Carbon. 1997;**35**(2): 259-266

[49] Mollinedo J, Schumacher JE, Chintala R. Influence of feedstocks and pyrolysis on biochar's capacity to modify soil water retention characteristics. Journal of Analytical and Applied Pyrolysis. 2015;**114**:100-108

[50] Van Zwieten L, Kimber S, Morris S, Chan KY, Downie A, Rust J, et al. Effect of biochar from slow pyrolysis of papermill waste on agronomic performance and soil fertility. Plant and Soil. 2010;**327**:235-246

[51] Zheng Z, Song-da Z, Ting-giang L, Feng-Liang Z, Zhen-Li H, He-ping Z, et al. Adsorption of ammonium phosphate from aqueous solution by biochar derived from phytoremediation of plants. Journal of Zhejiang University. Science. B. 2013;**14**(912):1152-1161

[52] Steiner C. Biochar carbon sequestration: Biorefinig Carbon Cycle Program. Athens, GA: University of Georgia; 2008

[53] Nwagwu C, Ede PN, Okoli IC, Chukwuka OK, Okoli CG, Moreki J. Effect of environmental factors and structural dimension of aerial pollution gas concentrations in tropical poultry pen in Nigeria. International Journal of Applied Poultry Research. 2012;**1**(1): 15-20

[54] Okoli IC, Anyaegbunam CN, Etuk EB, Uchegbu MC, Udedibie ABI. Socio-economic characteristics of poultry business entrepreneurs in Imo State, Nigeria. Journal of Agriculture and Social Research. 2004;**4**(2):100-111

[55] Charles EB. Litter management for confined turkeys. In: Poultry Science and Technical Guide. Vol. 41. The North Carolina Agricultural Extension Service Bulletin; 2005. pp. 3-7

[56] Sashikala M, Pillai GP, Xinguang R, Stuetz M. Odour abatement of poultry litter using odour control products. Chemical Engineering Transactions. 2012;**30**:247-257. Available from: www.acidic.it.cet

[57] Bhatti SA, Firkins JT. Kinetics of hydration and functional specific gravity of fibrous feed by-products. Journal of Animal Science. 1995;**73**(5): 1449-1458

[58] Kaske M, Hatiboglu S, Engelhadt WV. The influence of density and size of particles on rumination and passage from the reticulo-rumen of sheep. British Journal of Nutrition. 1992;**67**:235-244

[59] Lartey RB, Acquanh F, Nketia KS. Developing national capacity for manufacture of activated carbon from agricultural waste. The Ghana Engineering. 1999;**3**:45-50

[60] Tabbakh T, Barhoun R. Cleanup oil spills by activated carbons prepared from agricultural wastes. Indian Journal of Material Sciences. 2018;**16**(1):1-9

[61] Tumin ND, Luqman AC, Zawani Z, Suraya AR. Adsorption of copper from aqueous solution by *Elaisguineensis* kernel activated carbon. Journal of Engineering Science and Technology. 2008;**3**(2):180-189

[62] Tay JH, Chen XG, Jeyaseclan S, Graham N. Comparative study of anaerobically digested and undigested sewage sludge in preparation of activated carbons. Chemosphere. 2001; **44**:53-57

[63] Tsai WT, Lee MK, Chang YM. Fast pyrolysis of rice husk: product yields and composition. Bio-resource Technology. 2007;**98**:22-28

[64] Malik R, Ramtake DS, Water SR. Physico-chemical and surface characterization of adsorbent prepared from groundnut shell by $ZnCl_2$ activation and its ability to adsorb colour. Indian Journal of Chemical Technology. 2006;**13**:319-328

[65] Etuk IF, Ogbuewu IP, Iwuji TC, Okoli IC, Aladi NO, Williams E, et al. Physiological responses of broilers to drinking water from different sources in eastern Nigeria. International Journal of Agriculture and Rural Development. 2016;**19**(1):2422-2426

[66] Etuk IF, Ogbuewu IP, Okoli IC, Etuk EB, Iwuji TC, Obikoonu HO, et al. Quality of different water sources used in poultry and piggery farms in southeastern Nigeria. International Journal of Agriculture and Rural Development. 2014;**17**(2):1847-1852

[67] Widowati W, Asnah A. Biochars effect on potassium fertilizer and leaching potassium dosage for two corn-planting seasons. Agrivita. 2014;**36**: 165-171

[68] Hidayu AR, Muda N. Preparation of impregnated activated carbon from palm kernel shell and coconut shell for Co_2 capture. Procedia Engineering. 2016;**148**: 106-113

[69] Domingues RR, Trugilho PF, Silva CA, DeMelo ICNA, Melo LCA, Magriotis ZM, et al. Properties of biochar derived from wood and high-nutrient biomasses with the aim of agronomic and environmental benefits. PLoS One. 2017; **12**:e0176884

[70] EBC [European Biochar Certificate]. Guideline for a Sustainable Production of Biochar. Arbaz, Switzerland; 2012 Available from: http://www.european-biochar.org/en/download

[71] International Biochar Initiative. (2017). Standardized product definition and product testing guidelines for biochar that is used in soil. International Biochar Initiative, IBI-STD-1.1.

[72] Verheijen F, Diafas I, Jeffery S, Bastos A, Valde MVD. Biochar Application to Soils: A Critical Scientific Review of Effects on Soil Properties and Functions. Brussels: European Commission Joint Research Centre; 2010

[73] Martinez ML, Torres MM, Guzman CA, Maestri DM. Preparation and characterization of activated carbon from olive stones and walnut shells. Industrial Crop Production. 2006;**23**: 23-28

[74] Cagnon B, Py X, Guillot A, Stoecidi F, Chambar G. Contribution of hemicelluloses and lignin to the mass and the porous properties of chars and steam activated carbons from various lignocellulosic precursors. Bioresource Technology. 2009;**100**(1):292-298

[75] Campbell QP, Bunt JR, Kasaini H, Kruger DJ. The preparation of activated carbon from South Africa coal. Journal of the Southern African Institute of Mining and Metallurgy. 2012;**112**:37-44

[76] Abechi SE, Gimba CE, Uzairu A, Dalltu YA. Preparation and characterization of activated carbon from palm kernel shell by chemical activation. Resource Journal of Chemical Science. 2013;**3**(7):54-61

[77] Cetinkaya S, Sakintuna B, Yuyun Y. Formation of crystal structure during activated carbon production from

Turkish Elbistan lignite. Fuel and Chemical Division and Preparation. 2003;**48**(1):67-69

[78] Hirunpraditkoon S, Tunthong N, Ruangchai A, Nuithitku K. Adsorption capacities of activated carbons prepared from bamboo by KOH activation. World Academy of Science, Engineering and Technology. 2011;**78**:711-715

[79] Udoetok IA. Characterization of ash made from oil palm empty fruit bunches. International Journal of Environmental Sciences. 2012;**3**(1):518-524

[80] Kutlu HR, Unsal I, Gorgulu M. Effect of providing dietary wood (Oak) charcoal to broiler chicks and laying hens. Animal Feed Science Technology. 2001;**90**:213-226

[81] Evans AM, Loop SA, Moritz JS. Effect of poultry litter biochar diet inclusion on feed manufacture and 4–21-d broiler performance. Journal of Applied Poultry Research. 2015;**24**:380-386

[82] Jindal N, Mahipal SK, Mahajan NK. Toxicity of aflatoxin B_1 in broiler chicks and its reduction by activated charcoal. Research in Veterinary Science. 1994;**56**:37-40

[83] Honda K. Peripheral regulation of food intake in chickens: adiposity signals, satiety signals and others. World's Poultry Science Journal. 2021;**77**(2):301-312

[84] Sam AH, Troke RC, Tan TM, Bewick GA. The role of the gut/brain axis in modulating food intake. Neuropharmacology. 2012;**63**(1):46-56

[85] Tachibana T, Matsuda K, Kawamura M, Ueda H, Khan MSI, Cline MA. Feeding-suppressive mechanism of sulfated cholecystokinin (26-33) in chicks. Comparative Biochemistry and Physiology, Part A. Molecular and Integrative Physiology. 2012;**161**(4):372-378

[86] Wood SC. The control of food intake: Behavioral versus molecular perspectives. Cell Metabolism. 2009;**9**(6):489-498

[87] Dunn IC, Meddle ST, Wilson PW, Wardle AS, Law AS, Bishop C, et al. Decreased expression of the satiety signal receptor CCKAR is responsible for increased growth and body weight during domestication of chickens. American Journal of Physiology – Endocrinology and Metabolism. 2013;**304**(9):E909-E921

[88] Damron WS. Introduction to animal science, 4th edition. In: Pearson International Edition, S4 carlike publishing services. Edward Brothers and Phoenix: Colcord; 2009

[89] Irshad A. Effect of probiotics on broiler performance. International Journal of Poultry Science Science. 2006;**5**(6):593-597

[90] Neuvonen PJ, Kuusisto P, Vapaatalo H, Manninen V. Activated charcoal in the treatment of hypercholerolaemia: Dose-response relationships and comparison with cholestyramine. European Journal of Chemical Pharmacology. 1989;**37**:225-230

[91] Roosdiana A, Vidiastuti D, Herenda H. The preventive effect of activated charcoal on HDL levels and aorta histopathological profiles in hypercholesterol rat models. Journal of Physics: Conference Series. 2019;**1374**:012029

[92] Shabani A, Dastar B, Khomeiri M, Shabanpour B, Hassani S. Response of broiler chickens to different levels of

nanozeolite during experimental aflatoxicosis. Journal of Biological Sciences. 2010;**10**(4):362-367

[93] Majewska T, Zaborowski M. Charcoal in the nutrition of broiler chickens. Medycyna Weterynaryjina. 2003;**59**:81-83

[94] Iyaode II, Ibrahim HO, Uwade F, Shittu MW. Hematological and serum biochemistry of broilers strains (Cobbsand Arbor-acre) fed ginger. GSC Journal of Biological and Pharmaceutical Sciences. 2020;**11**(2):320-326

[95] Marcos BC, Fabricio PR, Hugo RM, Mara RBN, Antonio VM, Cristiane FPM. Biochemical blood parameters of broilers at different ages under thermo-neutral environment. World's Poultry Science Journal. 2012 Supplement 1, Expanded Abstract – poster presentation – chicken breeder and broiler production, 143

[96] Hidayat DA, Putra SS, Widiastuti E. Red Blood Profile of Broiler Chicken Fed with Tapioca Waste Fermented with *Acremoniumcharticola* and/or Antibiotic. UNS: National Seminar Department of Livestock; 2016

[97] Salam S, Sunarti D, Isroli. Physiological responses of blood and immune organs of broiler chicken fed dietary black cumin powder during dry seasons. JITAA. 2013;**38**(3): 185-191

[98] Sugiharto I, Widiastuti and Prabowo, N.S. Effect of turmeric extract on blood parameters, feed efficiency and abdominal fat content in broilers. JITAA. 2015;**36**(1):21-26

[99] Joseph V, Christopher AK, James VD, Ekta P, Richard FC. A retrospective review of the prehospital use of activated charcoal. The American Journal of Emergency Medicine. 2015; **33**(1):56-59

[100] Boonanuntanasarn S, Khaomek P, Pitasong T, Hua Y. The effects of the supplementation of activated charcoal on growth, health status and fillet composition-odor of Nile Tilapia (*Oreochromis niloticus*) before and after harvesting. Aquaculture International. 2014;**22**:1417-1436

[101] Ugbogo AE, Okezie E, Ijioma SN. Introducing Biochemistry Practical. Okigwe, Imo State, Nigeria: Justman Publishers International; 2017. pp. 143-202

[102] Niho MN, Mutoh M, Takahashi M, Tsutsumi K, Sugimura T, Wakabayashi K. Concurrent Suppression of Hyperlipidemia and Intestinal Polyp Formation by No-1886, Increasing Lipoprotein Lipase Activity in Min Mice. Tokyo: National Cancer Centre Research Institute; 2005

[103] Xenoulis PG, Steiner JMM. Lipid metabolism and hyperlipidemia in dogs. The Veterinary Journal. 2010;**183**:12-21

[104] Thrall MA. Hematologiae Bioquimica, Clinica Veterinaria. Sao Paulo: Roca; 2007 582 p

[105] Lumeji JT. Avian clinical biochemistry.In. In: Kaneko JJ, Harvey JW, Bruss ML, editors. Clinical Biochemistry of Domestic Animals. 5th ed. SanDiego, CA: Academy Press; 1997. p. 932

Chapter 9

Vermiconversion of Textile Industrial Sludge: Waste Management and Nutrients Recycling

Sharda Dhadse

Abstract

The present study aimed for the conversion of textile industrial sludge (TIS) amended with the cow dung into vermicompost operated by the epigenic earthworm *Eudrilus eugeniae.* To accomplish the intent of the experiment, the substrate was allowed to decompose for 30 days, under monitored environmental conditions. Three different combinations were prepared ($V_{25\%}$, $V_{50\%}$, and $V_{75\%}$) from TIS, and compared with V_{agro} (vermicompost prepared from agricultural waste) and V_{soil}. Among the entire three treatments, $V_{75\%}$ was shown by physicochemical parameters for *Trigonella foenum* (Fenugreek/Methi) plant growth, seed germination, and leave production in the tested pot. The maximum amount of available nitrogen, phosphorus, potassium (NPK) was recorded at $V_{75\%}$. On the other side, toxic metal (Cr, Mn, Cu, Pb Cd, and Zn) concentrations were diluted to minimum levels. The result advised that vermicomposting consider one of the alternative methods for waste management and energy recovery from industrial waste.

Keywords: textile industrial sludge, agricultural waste, toxic heavy metal, vermicomposting, energy recovery, waste management

1. Introduction

The increase in solid waste generation in developing countries is more worrisome than in developed countries owing to the shortage of supply and the need for suitable disposal techniques. The research was prepared the vermibed from three different vegetable waste, rice straw and cow dung in a different ratio, and inoculated with two different species of earthworm (*Eisenia fetida and Perionyx excavates*) *E. fetida* showed that the total organic carbon (TOC) decreased, pH was falling to neutral and the NPK ratio, microflora (nitrogen fixing bacteria (NFB), phosphate solubilizing bacteria (PSB), and total bacteria) was raised by vermicomposting [1]. Sludge generation in industries is generally a solid/semi-solid substance, which is contaminated with toxic elements and other contaminants. Generated waste management and recovery of energy is a difficult procedure. It is directly related to technological progress, economic growth, and industrialization. Various industries, such as pulp, paper, sugar, cement, tanneries produce sludge and disposal. To manage industrial waste, safe, eco-friendly, cost-effective, and socially acceptable techniques are needed [2]. Industrial sludge, irrespective of scale or location, presents a major challenge for all Indian industries as a result of strict national disposal requirements. Industrial growth has led to the higher production of goods, and enhanced urban livelihoods, but has degraded cities' environmental quality. The major issues faced due to toxic industrial wastes include: (a) groundwater and surface water pollution by the toxins in the sludge by leaching, (b) contamination caused by the heavy metals and chemicals present in the sludge, (c) color imparted on water or soil, (d) odor issue, and (e) risk to human health that calls for public concern. There have been several strategies for the disposal of solid waste worldwide. Open dumping, land-filling, composting, thermal drying, and incineration are some of the major techniques used for disposal of solid waste. Each approach has both benefits and disadvantages. There is no single approach with absolute applicability [3]. Industrial waste disposal methods are expensive and hard to handle, integrating the biosolid waste composting with vermistabilization provides a sustainable development value [4]. Industrial sludge disposal practicing on open roadsides/railway track beside and in agricultural fields in addition to poorly designed and maintained landfill sites by various Indian industries owing to high capital needs, operational costs for sludge treatment, and stricter waste disposal regulations, where there is a huge risk of soil or water contamination and health hazards eventually damage to the ecological balance. Also, open dumping is impractical as after a certain period there will be a limitation in open spaces. Because of the costs of installing sludge stabilization and dehydration systems, sludge management remains a challenging task [5].

Textile industrial sludge (TIS) comprises a mixture of organic and inorganic heavy metal complexes, such as Fe, Cu, Cd, Zn, and Cr. Textile industrial use various dyes and chemicals are employed in various steps and the emissions polluted with different inorganic and organic chemicals [6]. As compared to conventional methods of disposal methods, such as land-filling and incineration, vermicomposting is a better option ecologically and economically [7]. The industrial sludge stabilized by vermicomposting will reduce the toxic elements in the compost, moreover, it may apply in agricultural practices [8]. Agricultural solid residue can be converted and used for plant growth, it can provide nutrients and enhance the quality of the soil [9].

2. Methodology

2.1 Sample collection

Samples were collected from the textile industry where completely organic dyes are being used for coloring the bed materials. Therefore, the wastewater and sludge were containing organic waste. That helps to easily degrade the mechanism for earthworms.

2.2 Textile industrial sludge

It was collected from the textile industry, in Gujarat, India. It was allowed to air dry and converted to a fine powder. The main chemical characteristics were analyzed pH (8.12), TOC (15.7%), available nitrogen (890.6 kg/ha), available phosphorus (167.9 kg/ha), available potassium (3160.6 kg/ha), and C:N ratio as 394.8.

2.3 Cow dung

Raw cow dung was brought from a cow farm. Major properties of the cow dung were pH (8.12), TOC (16.6%), available nitrogen (752.6 kg/ha), available phosphorus (46.1 kg/ha), available potassium (2383.2 kg/ha), and the C:N ratio 494.

2.4 *Eudrilus eugeniae*

Healthy earthworms (*E. eugeniae)* were collected randomly from the vermiculture.

2.5 Stoichiometry

Sampled waste materials were dried under sunlight, dehydrated waste crushed into powder, and then poured into four different ports for decomposition.

2.6 Experimental design

Three feed mixtures had distinct ratios of TIS and cow dung, together one filled with only cow dung (CD). One-and-a-half-liter cylindrical mud vessels were lined with a layer of rice straw and packed with 600 g of crushed and air-dried CD in pot-1 (V_C), 450 g of crushed and air-dried CD with the combination of 150 g crushed and air-dried TIS in pot-2 ($V_{25\%}$), 300 g of crushed and air-dried CD and 300 g of crushed and air-dried TIS in pot-3 ($V_{50\%}$), and finally, 150 g of crushed and air-dried CD and 450 g of crushed and air-dried TIS were in the pot-4 ($V_{75\%}$) (**Table 1**). Vessels remained under darkness with room temperature 22–26°C and maintained moisture by 60–80% by sprinkling the required amount of water over the experiment period. To provide additional aeration, and overcome the volatile toxic chemicals the blend was tuned manually after 15 days we introduced 40 healthy earthworms (*E. eugeniae*) to the pods and coved the pods with a green mesh cloth to protect earthworms from rodents and heat. After 30 days decomposed mixture from the pod was taken out, separated the cocoons were allowed to be air-dried, finally, compost was grained and stored in plastic cylindrical pods for further applications.

2.7 Chemical analysis

Homogenized samples were collected from the reactor vessels and recorded the physicochemical parameters with different standard methodologies, pH and EC were recorded by pH conductivity meter, bulk density, porosity, and water holding capacity of vermicompost were taken and estimated in sediment [10]. Total available nitrogen, phosphorous, and potassium were estimated with the Kjeldahl method [11], Estimated the C:N ratio based on the measured quantity of C and N [12], TOC, and organic matter concentration (OMC) of the sample was recorded with the help of the titration method [13], heavy metals, such as Al, Ba, Cd, Co, Cr, Cu, Li, Mn, Mo, Ni, Pb, and Zn were estimated by the adaptation of atomic absorption spectrophotometer [14].

2.8 Seed germination and plant growth observations

Six pods were collected and filled with 250 g of soil and added 25 g of prepared vermicompost to V_C, $V_{25\%}$, $V_{50\%}$, and $V_{75\%}$, another cylinder V_{agro} has filled with prepared vermicompost, finally, the sixth pot V_{soil} was poured soil only it left without

Physico-chemical parameter	Treatment 1 (25% TIS)			Treatment 2 (50% TIS)			Treatment 3 (75% TIS)		
	Control	Inceptive	Eventual	Control	Inceptive	Eventual	Control	Inceptive	Eventual
pH	7.32 ± 043	7.82 ± 0.9	7.12 ± 1.01	6.24 ± 0.26	7.01 ± 0.45	6.97 ± 1.41	7.23 ± 1.11	6.98 ± 1.31	6.55 ± 1.28
EC (dS m^{-1})	690 ± 97	982 ± 91	1435 ± 37	728 ± 30	1107 ± 40	1967 ± 113	942 ± 118	1327 ± 103	2140 ± 73
Available nitrogen (kg/ha)	1354.7 ± 20.3	1402.1 ± 20	1517.8 ± 96.5	1312.6 ± 98.3	1498.7 ± 176.9	1618.1 ± 84	1526.8 ± 114.1	1672.1 ± 117.7	2101.8 ± 77.4
Available phosphorus (kg/ha)	254.1 ± 12	291.0 ± 69.9	362.2 ± 51.1	213.4 ± 193.42	289.7 ± 76.5	357.2 ± 11.1	268.6 ± 13.9	308.9 ± 21.4	383.5 ± 11.1
Available potassium (kg/ha)	4393.4 ± 204.2	4983.2 ± 848.8	6898.4 ± 755.2	4583.9 ± 449.3	5019.7 ± 87	8570.6 ± 136.4	5019.3 ± 996.1	7829.5 ± 108.2	10306.4 ± 178.4
Ca^{2+}%	8.0 ± 1	11.6 ± 1.5	15.2 ± 0.9	8.7 ± 0.8	10.2 ± 1.1	18.4 ± 1.2	8.7 ± 1.3	12.6 ± 1.2	27.2 ± 0.9
Mg^{2+}%	14.2 ± 1.1	15.1 ± 1	18.8 ± 1.3	14.6 ± 1.2	16.3 ± 0.4	20.4 ± 1	14.4 ± 1.3	17.9 ± 1.5	22.8 ± 1.4
Na^{+}%	1.22 ± 0.11	1.41 ± 0.11	2.54 ± 0.33	1.28 ± 0.16	1.41 ± 0.1	2.67 ± 0.68	1.26 ± 0.24	1.32 ± 0.05	2.97 ± 0.79
Bulk Density (g/cm^3)	0.64 ± 0.05	0.61 ± 0.07	0.51 ± 0.09	0.63 ± 0.11	0.62 ± 0.09	0.37 ± 0.08	0.64 ± 0.11	0.60 ± 0.09	0.25 ± 0.132
Porosity (%)	75.34 ± 5.22	77.92 ± 6.47	82.01 ± 2.01	75.26 ± 5.14	76.82 ± 6.7	90.32 ± 1.09	75.82 ± 0.85	77.26 ± 0.99	93.58 ± 1.34
TOC (%)	14.9 ± 1.1	14.2 ± 1.1	12.8 ± 1.2	14.7 ± 1.2	13.1 ± 0.8	10.08 ± 1.04	14.8 ± 1.4	09.2 ± 1.1	06.1 ± 0.9
C:N ratio	246.3 ± 32.1	241.1 ± 39	188.9 ± 14.3	245.8 ± 44.2	2440.5 ± 426.3	138.4 ± 13.2	246.7 ± 15.3	213.8 ± 12.3	65.00 ± 0.79

Note: All of the parameter values were mean values of five times repeated experiments.

Table 1.
Physico-chemical parameters with a standard deviation of TIS in different composition allowed to vermicompost (30 days).

Figure 1.
Effects of prepared vermicompost (textile sludge compost) on T. foenum germination, growth, and leaves production in the pot (Vagro, $V_{25\%}$, $V_{50\%}$, $V_{75\%}$, VC, and V_{soil}).

adding prepared vermicompost. After that in each pot, 30 soaked seeds of *Trigonella foenum* (Fenugreek/Methi) were sown and the six pots were monitored in regular intervals for seed germination, recording the aspects, such as the number of seeds germinated, the number of leaves produced, and measuring the plant height in 10 days frequent intervals of 30 days' time period (**Figure 1**).

3. Result and discussions

3.1 Mineral consignment of prepared vermicompost

Vermicompost prepared by various treatments on the expanse of feedstock after 30 days, the final product was more stable, odour free, appear dark brown, and enriched with nutrients. Newly formed vermicompost physicochemical parameter changes were recorded (**Table 1**). pH vermicompost was prepared with cotton industrial waste with a combination of sheep manure and the decomposition was carried by earthworms, they were observed that the pH was reduced. On the other side, compared with the vermicompost prepared without earthworms' treatment, cation exchange capacity (CEC), total mineralization was raised and total nitrogen was decreased but at the same time nitrates were raised in the prepared vermicompost [15]. pH plays an important role in vermicompost for encouraging plant growth, in this study, we observed the gradual reduction of pH in treatment 1–3 after 30 days of decomposition (T_1–T_3), the pH decreases due to the conversion of N, P, and organic material into nitrates, orthophosphates, and organic acids, it was helpful to identify the alkalinity nature. In this analysis, $V_{25\%}$ were recorded and exhibited maximum pH reduction from 7.82 to 7.12 with a standard deviation value of 0.9–1.01, the lowest reduction was noticed in $V_{50\%}$ 6.97 with a value of the standard deviation of 1.41, and the medium pH reduction was observed in $V_{75\%}$ treatment 6.55 with a standard deviation value of 1.28 (**Table 1**).

3.2 Electrical conductivity

EC measured based on the formation of TDS (total dissolved salts) in the decomposed substrate (vermicompost), it decreases in compost and vermiwash while decomposing time, moisture concentration in vermicomposting increases the electrical conductivity it can be observed in bio-fertilizers [16]. Increased electrical conductivity was observed in the decomposed vermicompost (after 30 days) due to the release of various minerals salts in available form, the highest EC was noticed in $V_{75\%}$ treatment 2140 dSm^{-1} with a standard deviation value of 73 and the lowest EC was recorded in $V_{50\%}$ 1967 dSm^{-1} with a standard deviation value of 113 (**Table 1**). Stabilized, textile sludge is a good source of nutrients, it contains various organic molecules and inorganic plant nutrients, which are essential for growth like NPK and many trace elements and can become a good fertilizer after vermistabilization free of chemicals and pathogens. It is an undesirable toxic bi-product from wastewater treatment plants and other industries; it can trigger biohazards in the environment [5].

3.3 Available nitrogen

Herbal and pharmaceutical effluents were exposed to vermitechnology, as the result, the wastewater and the herbal waste were converted to enriched nutrients [17, 18]. Yadav and Garg [2] demonstrated in their experiment, that bakery industrial sludge combined with cow dung generates valuable vermicompost, they set up the six plastic bins containing 100% CD + 0% bakery industry sludge (BIS) to 50% CD + 50% BIS and observed that all the bins showed a reduction in TOC, pH, and C:N ratio up to 65.4–83.5% but at same time increment was noticed in all bins. Maximum reduction in TOC and C:N ratio were observed in bin 1 in which the combination was 100% CD + 0% BIS and the highest increment in TKN (total Kjeldahl nitrogen) was in the bin in which the combination was 90% CD + 10% BIS and the highest increment in TAP (total available phosphorous) and TK (total potassium) content in bin 1 in which combination was 100% CD + 0% BIS and maximum biomass of worms were found in bin 6, which contain 50% CD + 50% BIS (**Table 1**). Utilization of sludge from recirculating aquaculture system (RAS) in vermicomposting and produced-mineral rich compost. They prepared the setup with 5%, 10%, 15%, and 20% RAS, respectively along with 200 g of shredded wheat straw with initial 70% moisture content and observed the percentage of RAS increased and an increased number of juvenile and cocoons were noticed. Moreover, the end product of this sludge holed a higher amount of available nitrogen, available phosphorous, and other minerals [19]. Available nitrogen was increased in decomposed vermicompost, it was observed that $V_{75\%}$ treatment raised from 1672.1 to 2101.8 kg/ha with a standard deviation value of 117–77, and the lowest available nitrogen concentrations were recorded in $V_{25\%}$ 1517.8 kg/ha with a standard deviation value of 96.5 (**Table 1**). Vermicompost from the sewage sludge along with cow dung, they set up the 4-treatment contained sewage sludge and cow dung in ratios 70:30, 80:20, 90:10, and 95:5. Treatment ratio contained 70:30 (SS:CD) and 80:20 (SS:CD) observed the highest survival and reproduction rate and in ratio 95:05 any earthworm did not survive and in ratio 90:10 observed the highest available nitrogen, available phosphorous and other minerals [20].

3.4 Available phosphorus

Biofertilizer was prepared from municipal sewage sludge (MSS) through the vermicomposting process using tiger worms (*E. fetida*) after 21 days of vermicomposting process available nitrogen was increased up to 19.6–35.7 mg/l, total phosphorous from 9.45 to 10.87 mg/l and TP from 3.44 to 4.80 mg/l, and conversion of MSS to vermicompost found to be 93% by weight and worm biomass showed 30% increment from its initial weight [21]. Mineralization and mobilization of phosphorus by bacterial and phosphatase activity of earthworms could be the main reason for phosphorus improvement in vermicompost. Maximum available phosphorus was recorded in $V_{75\%}$ 383.5 kg/ha with a standard deviation value of 11.1, and the lowest amount of available phosphorus was noticed in $V_{50\%}$ 357.2 kg/ha with a standard deviation value of 1.11 (**Table 1**). Textile mill sludge (TMS) along with cow dung (CD) can be utilized as the raw material for vermicompost, they inoculated three microbial species in their experiment and found out that as the cow dung concentration decreased the growth of microbes was also decreased but as the inoculation of microbes in vermibed increased, available nitrogen and available phosphate were increased comparison to which did not have any microbes inoculated in them, they also noticed that vermicompost which only contains CD was more productive for the growth and multiplication of all the three bacteria than CD + TMS vermicompost, *Azotobacter chroococcum* treated vermicomposts showed the maximum available nitrogen, *Pseudomonas maltophila* inoculated CD vermicompost showed the maximum available phosphorous and finalized that as the number of microbes increased during vermicomposting the available N, P also increased and C:N ratio also decreased [22].

3.5 Available potassium

Vermicomposting was carried out by utilizing a different variety of waste, such as textile sludge, agricultural residue, and vegetable waste, final compost was shown an increase in phosphorous (1.4–6.5 folds) and potassium (4.4–5.8 folds) concentrations in the feed mixture [23]. Available potassium was observed to be increased in prepared vermicompost (after 30 decomposition) due to liberation of different soluble mineral salts in organic matter decomposition, and the potassium mineral salts were present in the form of available. An increased amount of available potassium was noticed in all of the treatments among all of them $V_{50\%}$ was raised at a high concentration of 8570.6 kg/ha with a standard deviation value of 136.4 and the lowest increase was noticed in $V_{25\%}$ 6898.4 kg/ha with a standard deviation value of 755.2 (**Table 1**). Vermicompost was prepared from pig manure with dissolved organic matter and observed the effects on heavy metal behavior. Pig manure mixed with rice straw in different combinations. Concentrations of Cu and Zn in earthworms increased from 8.24 and 17.63 to 40.75 and 362.78 mg/kg separately after vermicomposting, and also increased their availability, the C:N ratio also decreased after vermicomposting from 10.37 to 8.60. The available NPK was observed to be increased after vermiconversion of pig manure with rice straws [24].

3.6 Calcium (Ca^{2+}), magnesium (Mg^{2+}), and sodium (Na^{+})

Vermiconversion of paper mill sludge by the earthworms drives the sludge into mineralization effectively and converts the bound form into free minerals forms, Ca, Mg, and Na concentrations were found to be more (12.9%) in treatment 2 [25]. Ca^{2+}, Mg^{2+}, and Na^{+} concentrations were increased in prepared textile vermicompost, among all the three treatments $V_{75\%}$ reactor recorded as highest concentrations of 27.2%, 22.8%, and 2.97% with standard deviation values of 0.9, 1.4, and 0.79, respectively, lowest concentrations were found in $V_{25\%}$ as 15.2%, 18.8%, and 2.54 with standard deviation values of 0.9, 1.3, and 0.33, respectively. Calcium concentration was more than the other elements in the prepared vermicompost, and earthworms (*E. eugeniae*) operated the mineralization effectively and converted a huge amount of Ca^{2+}, Mg^{2+}, and Na^{+} in the organic matter (**Table 1**). Vermicomposting of bakery waste and cow dung by employing the earthworms *E. fetida*, thus the resulting nutrient-rich vermicompost produced, it was highly stabilized and mineralized than the initial food sludge waste more over heavy metal concentration were raise was observed in the newly developed vermicompost [2].

3.7 Bulk density and porosity

It has been found that concentrations of the minerals were more in aquatic weeds than in prepared vermicompost, according to the Fertilizer Control Order [26] decrease in bulk density was due to the gut action performed inside the earthworm and it caused the particle size changes. The bulk density and porosity of the newly prepared vermicompost were analyzed and recorded, it was observed that the $V_{75\%}$ treatment contain less bulk density of 0.25 g/cm^3 with a standard deviation value of 0.132, and the highest porosity of 93.58% with a standard deviation value of 1.34, then the resulting water holding capacity of the prepared compost was more. $V_{50\%}$ treatment noticed a medium bulk density of 0.37 g/cm^3 with a standard deviation value of 0.08 and a medium porosity of 90.32% with a standard deviation value of 1.09, a higher bulk density (0.51 g/cm^3) with a standard deviation value of 0.09, and lowest porosity of 82.01% with a standard deviation value of 2.01 was recorded in the $V_{25\%}$ treatment (**Table 1**).

3.8 Total organic carbon

A study was conducted on food industrial sludge combined with various organic waste and allowed for decomposition and the final results were an increase in total nitrogen, phosphorous, sodium, and potassium at the same time decrease in pH, TOC, and C:N ratio was noticed [27]. TOC was observed in prepared vermicompost, the result suggested that a useful biodegradable pool of organic carbon was slowly used during the reduction of TOCs. In $V_{75\%}$, the maximum reduction was noticed at 06.1% with a standard deviation value of 0.9, and the lowest amount of TOC was noticed in $V_{25\%}$ treatment at 12.8% with a standard deviation value of 1.2, the loss was due to the utilization of carbon by earthworms and microbial consumption and the microbial respiration leads to loss of carbon in the form of CO_2 during the decomposition. Further, the rise in earthworms' population, due to the conversion of some part of the organic fraction of the substrate, can also cause the stabilization of organic matter by earthworms. The lowest TOC content indicates the richness of humic substances, stability, and maturity of compost (**Table 1**). The key concerns related to conventional thermophilic composting are the process takes a long period, the pace of turning of the waste, the size and volume of the materials are often needed to be decreased to provide the necessary surface area, and the loss of nutrients during the lengthy process and the final product is heterogeneous nature. In this composting process, to maintain aerobic conditions, the waste must be turned regularly or aerated in some other way. Mostly, this requires powerful and costly machinery to handle the residuals as efficiently as necessary on a massive scale [4].

3.9 C:N ratio

Vermicompost prepared from milk processing industrial sludge combined with sugarcane trash and cow dung. They prepared nine various combinations of vermibeds with MPIS, ST, and CD. MPIS (60%) + CD (10%) + ST (30%) and MPIS (60%) + CD (10%) + WS (30%) containing mixture show highest reduction, organic carbon and C:N ratio and it exhibited highest raised concentrations in available nitrogen, available phosphorous, and exchangeable potassium [28]. C:N ratios minimized with time in all the vermicomposting treatments, the decline in the C:N ratio may be due to the loss of carbon through microbial respiration in the form of CO_2. In the $V_{75\%}$ treatment, the maximum reduction of C:N ratio (65.0) with a standard deviation value of 0.79 was recorded and the lowest C:N ratio was noticed in $V_{25\%}$188.9 with a standard deviation value of 14.3 (**Table 1**). The research observed the N and P content after inoculation of *A. chroococcum* strains, *Azospirillum lipoferum*, and the phosphate solubilizing *Pseudomonas striata.* Total six treatments were prepared, such as T1-(V + *A. chroococcum*. Mac 27), T2-(V + *A. chroococcum* .54 – 1), T3-(V + *A. chroococcum* .35-47), T4-(V + *A. lipoferum*), T5-(V + *P. striata*), and T6-(V + *P. striata* + Mussoorie rock phosphate 1%). On day 0, the vermicompost initially contained only 1.40 (g/100 g) of N. On the 60th day following inoculation with *A. chrooccocum* (Mac 27), it was increased to 2.72 (g/100 g) Chrooccal (Mac 27). Similarly, N content increased to 2.53 and 2.50 (g/100 g) with inoculation of other Azotobacter strains. *P. striata* also increase the phosphorous content after inoculation in vermicompost [29].

3.10 Heavy metal concentration in vermicompost

Aquatic weeds accumulated with most of the essential elements can be used in the food chain, the paper deals with bioconversion of textile sludge decomposition with the help of earthworm feeding. Heavy metal concentrations of the textile sludge were decreased by the action of earthworm digitations. Significant toxic element (Cd, Cr, Ni, Cu, Pb, and Zn) reduction was observed in the co-vermistabilization experiment [30]. Heavy metal degradation in any substance is not possible but it can be reduced by implementing the recommended methodologies, which are immobilization and toxic reduction/removal. The critical reciprocity of the earth warms and lowers the concentrations of heavy metals in developed vermicompost. Maximum heavy metal concentrations in the developed vermicompost were aluminum (Al), barium (Ba), lead (Pb), and zinc (Zn) (**Table 2**). Significant reduction of Al had been observed in $V_{75\%}$ treatment, it was from 5.01 to 2.61 ppm with the values of standard deviation 0.02–0.36 by the earthworms (*E. eugeniae*), and less reduction of Al had been noticed in $V_{25\%}$ treatment, it was from 3.56 to 2.24 ppm with the values of standard deviation 0.08–0.1. The highest barium content reduction was recorded in the $V_{75\%}$ treatment, which was from 30.56 to 3.55 ppm with the values of standard deviation 1.11–0.56 and the lowest reduction capacity was sported in the $V_{25\%}$ reactor, which was from 11.79 to 1.57 ppm with the values of standard deviation 0.8–0.1. Cd, Co, Cr, Cu, Fe, Li, and Ni concentrations were less in all of the treatment reactors (**Table 2**). Manganese (Mn) reduction was more in the $V_{75\%}$ reactor, it was minimized from 1.47 to 0.05 ppm, with the values of standard deviation 0.06–0.01. Lead (Pb) content was less in the $V_{75\%}$ reactor and the zinc minimization and presence were good in the $V_{75\%}$ reactor, it was from 0.02 to 0.01 ppm with the values of standard deviation 0.001–0.079 (**Table 2**). An experiment was carried out to immobilize the heavy metals in fresh industrial sludge and composite industrial sludge and added the prepared compost to degraded agricultural soils, a surprisingly significant decrease in the heavy metals in the agricultural subsoils was noticed [31].

Heavy metals	Treatment 1 (25% TIS)			Treatment 2 (50% TIS)			Treatment 3 (75% TIS)		
	Control	Inceptive	Eventual	Control	Inceptive	Eventual	Control	Inceptive	Eventual
Al	5.55 ± 0.88	3.56 ± 0.08	2.24 ± 0.1	5.55 ± 0.88	4.95 ± 0.51	2.48 ± 0.24	5.55 ± 0.88	5.01 ± 0.02	2.61 ± 0.36
Ba	37.35 ± 2.12	11.79 ± 0.8	1.57 ± 0.1	37.35 ± 2.12	21.15 ± 1.04	2.36 ± 0.15	37.35 ± 2.12	30.56 ± 1.11	3.55 ± 0.56
Cd	0.0004 ± 0.0001	0.0002 ± 0.0003	0.0002 ± 0.0003	0.0004 ± 0.0001	0.0009 ± 0.0001	0.0002 ± 0.0003	0.0004 ± 0.0001	0.0003 ± 0.0001	0.0003 ± 0.0001
Co	0.021 ± 0.001	0.02 ± 0.01	0.011 ± 0.079	0.021 ± 0.001	0.02 ± 0.01	0.01 ± 0.01	0.02 ± 0.001	0.02 ± 0.001	0.01 ± 0.001
Cr	0.009 ± 0.001	0.004 ± 0.001	0.003 ± 0.005	0.009 ± 0.001	0.007 ± 0.001	0.003 ± 0.005	0.009 ± 0.001	0.008 ± 0.001	0.004 ± 0.001
Cu	0.083 ± 0.003	0.06 ± 0.01	0.045 ± 0.004	0.08 ± 0.003	0.06 ± 0.01	0.06 ± 0.01	0.08 ± 0.003	0.07 ± 0.01	0.10 ± 0.001
Fe	0.103 ± 0.001	0.1 ± 0.8	0.008 ± 0.001	0.10 ± 0.001	0.1 ± 0.8	0.04 ± 0.01	0.10 ± 0.001	0.1 ± 0.8	0.06 ± 0.01
Li	0.009 ± 0.001	0.003 ± 0.005	0.001 ± 0.001	0.009 ± 0.001	0.007 ± 0.001	0.002 ± 0.001	0.009 ± 0.001	0.007 ± 0.001	0.003 ± 0.005
Mn	1.55 ± 0.04	0.95 ± 0.04	0.02 ± 0.01	1.55 ± 0.04	1.1 ± 0.2	0.04 ± 0.01	1.55 ± 0.04	1.47 ± 0.06	0.05 ± 0.01
Ni	0.04 ± 0.01	0.02 ± 0.01	0.01 ± 0.01	0.04 ± 0.01	0.03 ± 0.01	0.01 ± 0.01	0.04 ± 0.01	0.034 ± 0.005	0.016 ± 0.001
Pb	0.11 ± 0.79	0.10 ± 0.8	0.007 ± 0.001	0.11 ± 0.79	0.1 ± 0.8	0.007 ± 0.001	0.11 ± 0.79	0.09 ± 0.01	0.008 ± 0.001
Zn	0.03 ± 0.01	0.02 ± 0.01	0.02 ± 0.01	0.03 ± 0.01	0.01 ± 0.01	0.01 ± 0.01	0.03 ± 0.01	0.02 ± 0.001	0.011 ± 0.079

Note: All of the parameter values were mean values of five times repeated experiments.

Table 2.
Concentration of heavy metals with standard deviation of TLS after vermitreatment (30 days).

3.11 Growth rate estimation of *T. foenum* (Fenugreek/Methi) seeds in soil mixed with V_C, $V_{25\%}$, $V_{50\%}$, $V_{75\%}$ prepared vermicompost

An experiment carried out on the combination of vermicompost and NPK fertilizers, enhances the yield of *Allium sativum*, six experimental plots were prepared: T1 (recommended NPK), T2 (vermicompost 15 t/ha), T3 (20 t/ha vermicompost), T4 (15 t/ha vermicompost + 50% NPK), T5 (15 t/ha farmyard manure), and T6 (farmyard manure 15 t/ha + 100% recommended NPK) to observed the effect on garlic, it was exhibited the highest growth in root length, shoot length leaf length in T4 treatment plot [32]. A total of six cylindrical plastic pots were collected and filled with different portions of prepared compost, cow dung, and soil, one reactor total portion was filled with soil and labeled as V_{soil}, second one was filled with cow dung, labeled as V_C reactor and the same three cylinders were ($V_{25\%}$, $V_{50\%}$, and $V_{75\%}$) filled with approximately 250 g of soil in each of them 25 g of prepared 25%, 50%, and 75% vermicompost was poured in different reactors and labeled as V_C, $V_{25\%}$, $V_{50\%}$, and $V_{75\%}$. Every reactor was sown with 30 seeds of *T. foenum* (Fenugreek/Methi) plant, and all of the five reactors were monitored for seed germination, the number of leaves production, and the growth of plants in 10 days frequent intervals of a total of 30 days (**Figure 1**). An experiment was conducted on the chemical nutritional variations in vermicompost, and pit compost, it was tested on the growth of *Pisum sativum* and observed the pot composed of vermicompost. It was exhibited that the maximum growth in every parameter of *P. sativum* plant as compared to pit compost, and garden soil (control). They noticed significant growth of root length, more number of leaves production, and healthy growth of plant height in the vermicompost, the above parameters were less when compared to the control and pit compost [33]. In V_{agro} pot 28 seeds were geminated out of 30 sown seeds, $V_C$cow dung pot was observed with 26 seeds germination out of 30 sown seeds, and $V_{25\%}$, $V_{50\%}$, and $V_{75\%}$ pots were observed with 26, 27, and 29 germinated seeds out of 30 sown seeds, V_{soil} was observed with 25 seed germination. Maximum seed germination was noticed in the $V_{75\%}$ reactor due presence of 75% of the prepared vermicompost in the sampled pot soil and the minimum seed germination was observed in V_C and $V_{25\%}$ pot due to the low nutritional value (**Figure 2**). Pig manure-based vermicompost when used for the growth

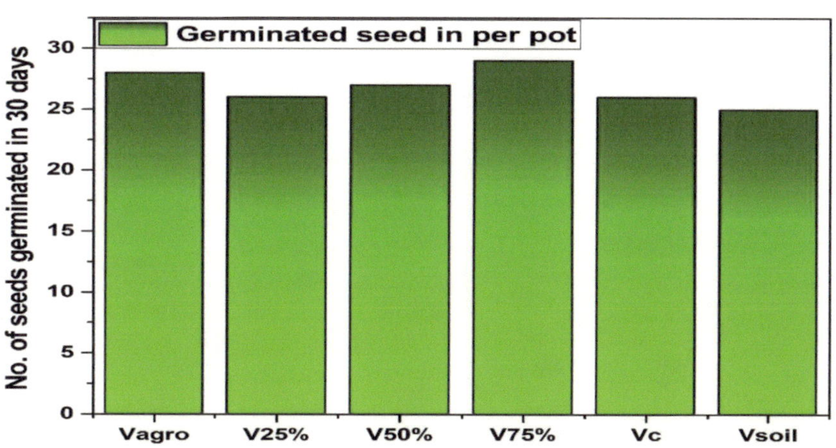

Figure 2.
Number of T. foenum plant seeds germinated on prepared textile sludge vermicompost.

of tomato (*Lycopersicon esculentum* Mill.). It was observed that tomato seedlings had been decreased in the potting mixture containing 100% pig manure vermicompost, likely due to high soluble salt concentrations in vermicompost and lower porosity and aeration. The growth of tomato seedlings was maximum after metro-mix 360 was replaced with between 25% and 50% pig manure vermicompost, with better growth occurring in combinations of pig manure vermicompost and regularly treated with a liquid fertilizer solution than in those without fertilizer [34].

3.12 Plant growth in relation with height

The relative rates of growth (RGR), based on the primary data, net assimilation rate (NAR), leaf area ratio (LAR), and components, thereof, specific leaf area (SLA) and leaf weight fraction (LWF) was calculated for the nursery stage and the transplant date, respectively. The growth response coefficients are based on the assumption RGR = NAR × SLA × LWF (GRC), the relative contribution of each parameter to an RGR change was calculated for NAR, SLA, and LWF. Vermin compost was discovered an effective growth medium for the propagation of vegetable seedlings, used individually or in the mixture [35]. The mean plant growth (height) was significant in $V_{75\%}$ pot, grown was observed up to 15.7 cm in 30 days, it was due to the availability of nutrients in prepared vermicompost from textile sludge. On the other hand, lesser plant height was observed in V_{soil} which was grown up to 11.3 cm. V_{soil} pot was filled with soil only, and no additional vermicompost/nutrients were available in the pot, the resulting in the lesser plant growth (**Figure 3**).

3.13 Mature plant leaves production and leaf growth

The highest growth was observed in plants treated with humic acid-rich vermicompost, which was prepared using fungal pretreatment. The highest root and shoot weight were also observed in plants treated with HARV, as compared to normal vermicompost and control (without compost), HARV treated plants observed 109.17%

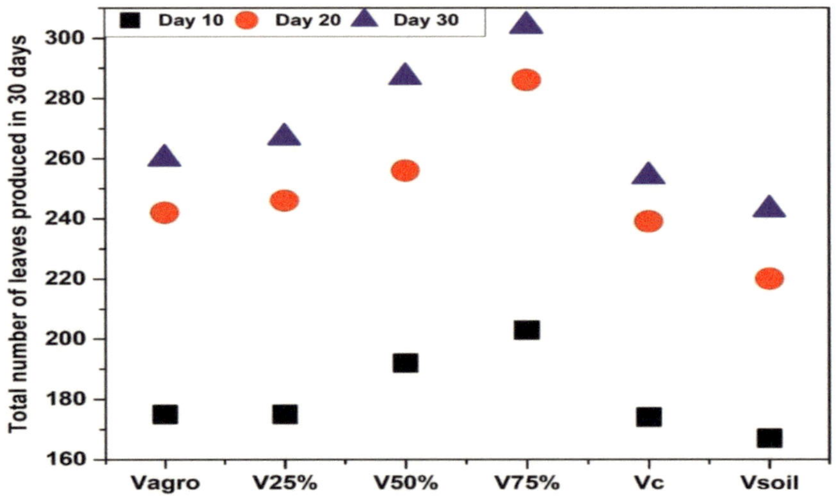

Figure 3.
T. foenum plant growth (height cm) by textile sludge vermicompost.

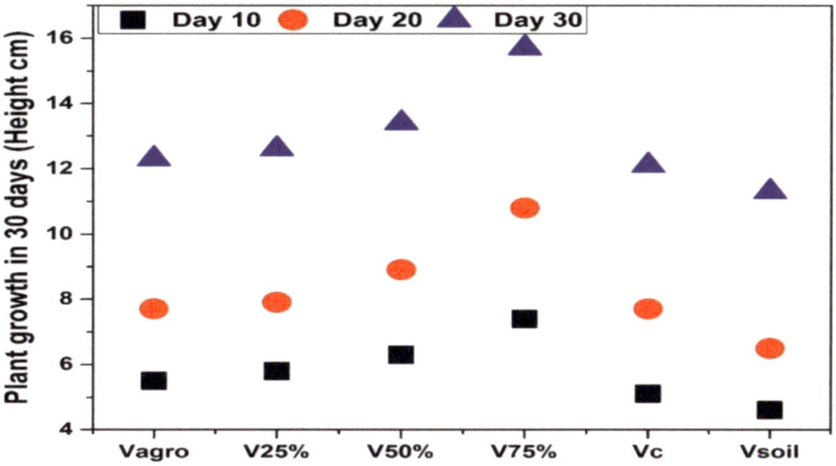

Figure 4.
T. foenum plant leaves production by textile sludge vermicompost.

plant yield, 82.97% in root biomass, and 51.61% in total height as compared to control in which any kind of vermicompost was not used [36]. Matured leaves were counted per pot in all of the five sown pots, the vermicompost use was not significantly different but in the absence of vermicompost noticed the difference in the formation of leaf number count. It was significantly more in all the vermicompost used media as compared to the control soil media. Height mature leaf count was found in $V_{75\%}$ pot (304 leaves) and the lowest count was found in V_{soil} pot (243 leaves) (**Figure 4**). A research team conducted two experiments in the greenhouse to observe the effect of peat compost and vermicompost on the growth of *Sorghum bicolor.* Two types of compost were used, pig manure vermicompost and pet compost, the sterile seed of *Sorghum bicolor* was grown in one experiment containing pet compost and experiment 2 contain vermicompost. After 21 days of germination, they were observed that pod compost experiment 1 induced root colonization in the plant after inoculation of AMF, and the dry weight was more in plant treated with vermicompost but did not induce any root colonization in plants [37].

4. Conclusion

TIS had significant organic and inorganic nutritional value with very low handling costs the disposal management problems can be overcome. The TIS waste can also be used in energy and nutrient recovery rather than used for landfill. Management and energy recovery from the TIS mixed with cow dung in different compositions was attempted to vermicompost by employing the earthworms. The final compost matter found was nutrient-rich, free from odor, it was stable, and highly mature, among all of the treatments $V_{75\%}$ reported the highest NPK values and micronutrients (Ca^{+2}, Mg^{+2}, and Na^{+2}) for plant growth.

Acknowledgements

The authors are thankful to the CSIR-NEERI for supporting the research.

Author details

Sharda Dhadse
Environmental Biotechnology and Genomics Division (EBGD), CSIR-National Environmental Engineering Research Institute (NEERI), Nagpur, India

*Address all correspondence to: sharda.dhadse@gmail.com

References

[1] Hussain N, Singh A, Saha S, Kumar MVS, Bhattacharyya P, Bhattacharya SS. Excellent N-fixing and P-solubilizing traits in earthworm gut-isolated bacteria: A vermicompost based assessment with vegetable market waste and rice straw feed mixtures. Bioresource Technology. 2016;**222**:165-174

[2] Yadav A, Garg VK. Biotransformation of bakery industry sludge into valuable product using vermicomposting. Bioresource Technology. 2019; **274**:512-517

[3] Kumar GV, Renuka G, Anoop Y. Potential of vermicomposting technology in solid waste management. In: Current Developments in Solid-state Fermentation. New York, NY: Springer; 2008. pp. 468-511

[4] Ndegwa PM, Thompson SA. Integrating composting and vermicomposting in the treatment and bioconversion of biosolids. Bioresource Technology. 2001;**76**(2):107-112

[5] Sinha RK, Herat S, Bharambe G, Brahambhatt A. Vermistabilization of sewage sludge (biosolids) by earthworms: Converting a potential biohazard destined for landfill disposal into a pathogen-free, nutritive and safe biofertilizer for farms. Waste Management & Research. 2010;**28**(10):872-881

[6] Guha AK, Rahman O, Das S, Hossain S. Characterization and composting of textile sludge. Resources and Environment. 2015;**5**(2):53-58

[7] Mowla D, Tran HN, Allen DG. A review of the properties of biosludge and its relevance to enhanced dewatering processes. Biomass and Bioenergy. 2013;**58**:365-378

[8] Yadav A, Suthar S, Garg VK. Dynamics of microbiological parameters, enzymatic activities and worm biomass production during vermicomposting of effluent treatment plant sludge of bakery industry. Environmental Science and Pollution Research. 2015;**22**(19):14702-14709

[9] Yadav A, Garg VK. Recycling of organic wastes by employing *Eisenia fetida*. Bioresource Technology. 2011;**102**(3):2874-2880

[10] Coutts JRH. "Single Value" soil properties. A study of the significance of certain soil constants. III. Note on the technique of the Keen-Raczkowski box experiment (with three text-figures.). Journal of Agricultural Science. 1930;**20**(3):407-413

[11] Jackson ML. Soil Chemical Analysis. New Delhi: Prentice Hall of India Ltd.; 1973. pp. 219-221

[12] Schoenau JJ, O'Halloran IP. Sodium bicarbonate-extractable phosphorus. In: Carter MR, Gregorich EG, editors. Soil Sampling and Methods of Analysis. CRC Press; 2007

[13] Walkley A, Black IA. An examination of the Degtjareff method for determining soil organic matter, and a proposed modification of the chromic acid titration method. Soil Science. 1934;**37**(1):29-38

[14] Kumar MS, Rajiv P, Rajeshwari S, Venckatesh R. Spectroscopic analysis of vermicompost for determination of nutritional quality. Spectrochimica Acta Part A: Molecular and Biomolecular Spectroscopy. 2015;**135**:252-255

[15] Albanell E, Plaixats J, Cabrero T. Chemical changes during vermicomposting (*Eisenia fetida*) of sheep manure mixed with cotton industrial wastes. Biology and Fertility of Soils. 1988;**6**(3):266-269

[16] Musaida M, Phiri A, Chirinda N, Muredzi P, Govhaand J, Sengudzwa T. Vermicomposting of waste corn pulp blended with cow dung manure using Eisenia fetida. International Journal of Chemical, Molecular, Nuclear, Materials and Metallurgical Engineering. 2012;**6**(8):1080

[17] Dhadse S, Chaudhari PR, Satyanarayan S, Wate SR. Vermitreatment of pharmaceutical wastewaters and nutrient bioassay of treated effluents for reuse as irrigation water. American Journal of Engineering Research (AJER). 2014;**3**(8):113-123

[18] Dhadse S, Satyanarayan S, Chaudhari PR, Wate SR. Vermifilters: A tool for aerobic biological treatment of herbal pharmaceutical wastewater. Water Science and Technology. May 2010;**61**(9):2375-2380

[19] Kouba A, Lunda R, Hlaváč D, Kuklina I, Hamáčková J, Randák T, et al. Vermicomposting of sludge from recirculating aquaculture system using Eisenia andrei: Technological feasibility and quality assessment of end-products. Journal of Cleaner Production. 10 Mar 2018;**177**:665-673

[20] Ludibeth SM, Marina IE, Vicenta EM. Vermicomposting of sewage sludge: Earthworm population and agronomic advantages. Compost Science & Utilization. 1 Jan 2012;**20**(1):11-17

[21] Zularisam AW, Zahirah ZS, Zakaria I, Syukri MM, Anwar A, Sakinah M. Production of biofertilizer from vermicomposting process of municipal sewage sludge. Journal of Applied Sciences. 2010;**10**(7):580-584

[22] Kaushik P, Yadav YK, Dilbaghi N, Garg VK. Enrichment of vermicomposts prepared from cow dung spiked solid textile mill sludge using nitrogen fixing and phosphate solubilizing bacteria. The Environmentalist. 2008;**28**(3):283-287

[23] Garg P, Gupta A, Satya S. Vermicomposting of different types of waste using *Eisenia foetida*: A comparative study. Bioresource Technology. 2006;**97**(3):391-395

[24] Zhu W, Yao W, Zhang Z, Wu Y. Heavy metal behavior and dissolved organic matter (DOM) characterization of vermicomposted pig manure amended with rice straw. Environmental Science and Pollution Research. 2014;**21**(22):12684-12692

[25] Natarajan N, Gajendran M. Vermiconversion of paper mill sludge for recycling the nutrients using earthworm *Eudrilus eugeniae*. International organization of scientific research. Journal of Environmental Science, Toxicology and Food Technology. 2014;**8**(9):06

[26] FCO. The Fertilizer (Control) Order, 1985 and the Essential Commodities Act, 1955. 1985

[27] Garg VK, Suthar S, Yadav A. Management of food industry waste employing vermicomposting technology. Bioresource Technology. 1 Dec 2012;**126**:437-443

[28] Suthar S, Mutiyar PK, Singh S. Vermicomposting of milk processing industry sludge spiked with plant wastes. Bioresource Technology. 1 Jul 2012;**116**:214-219

[29] Hemati A, Alikhani HA, Pourbabaee AA, Bagheri Marandi G. Enriching vermicompost by nitrogen fixing and phosphate solubilizing bacteria. Journal of Agricultural Science and Sustainable Production. 6 Nov 2013;**23**(3):117-128

[30] Dhadse S, Alam SN, Rao MM. Development of nutrient rich biofertilizer by co-vermistabilization of aquatic weeds using herbal pharmaceutical wastewater along with sediment of lake. Bioresource Technology Reports. 1 Feb 2021;**13**:100633

[31] Arif MS, Riaz M, Shahzad SM, Yasmeen T, Ashraf M, Siddique M, et al. Fresh and composted industrial sludge restore soil functions in surface soil of degraded agricultural land. Science of the Total Environment. 1 Apr 2018;**619**:517-527

[32] Suthar S. Impact of vermicompost and composted farmyard manure on growth and yield of garlic (*Allium stivum* L.) field crop. 2009

[33] Khan A, Ishaq F. Chemical nutrient analysis of different composts (Vermicompost and Pitcompost) and their effect on the growth of a vegetative crop *Pisum sativum*. Asian Journal of Plant Science and Research. 2011;**1**(1):116-130

[34] Atiyeh RM, Edwards CA, Subler S, Metzger JD. Pig manure vermicompost as a component of a horticultural bedding plant medium: Effects on physicochemical properties and plant growth. Bioresource Technology. 2001;**78**(1):11-20

[35] Kaciu S, Babaj I, Sallaku G, Balliu A. The influence of vermicompost on plant growth characteristics and stand establishment rate of pepper (*Capsicum annuum* L.) seedlings under saline conditions. Journal of Food, Agriculture & Environment. 2011;**9**(1):488-490

[36] Maji D, Misra P, Singh S, Kalra A. Humic acid rich vermicompost promotes plant growth by improving microbial community structure of soil as well as root nodulation and mycorrhizal colonization in the roots of *Pisum sativum*. Applied Soil Ecology. 2017;**110**:97-108

[37] Cavender ND, Atiyeh RM, Knee M. Vermicompost stimulates mycorrhizal colonization of roots of Sorghum bicolor at the expense of plant growth. Pedobiologia. 2003;**47**(1):85-89